Lecture Notes in Mobility

Series Editor

Gereon Meyer, VDI/VDE Innovation und Technik GmbH, Berlin, Germany

Lecture Notes in Mobility (LNMOB) is a book series that reports on the latest advances in research, development and innovations for the intelligent, connected and sustainable transportation systems of the future, comprising e.g.:

- Electric and hybrid vehicles
- Energy-efficient vehicles
- Alternative and optimized powertrains
- Vehicle automation and driver assistance
- Clean and intelligent transportation systems
- Interfaces with transportation networks and power grids
- Mobility services
- Business models
- Public policy

LNMOB publishes edited conference proceedings, contributed volumes and monographs that present firsthand information on cutting-edge research results and pioneering innovations as well as new perspectives on classical fields, while maintaining Springer's high standards of excellence. Also considered for publication are dissertations and other related material of exceptionally high quality and interest. The subject matter should be original and timely, reporting on the latest developments in all areas covered by this series. Concerns about climate change, air quality and energy security have sparked a paradigm shift regarding the transportation of passengers and goods. Vehicle technologies are now changing radically towards higher energy efficiency and the use of alternative fuels. At the same time, transportation is increasingly managed in a smart manner, connecting a multitude of modes and making use of a variety of energy sources. Moreover, vehicles are becoming more and more assistive and autonomous, which will lead to even higher efficiency, fewer fatalities and more convenience. These developments are to be supported by novel services, processes and value creation, as well as by suitable public policies. LNMOB identifies the most promising conceptual work, groundbreaking findings and innovations from the industrial engineers and academic experts in this field. The overall aim of this series is the exchange of ideas to strengthen the scientific community, to create new value chains and to accelerate the move towards sustainable transportation systems (road, rail, water- and airborne) around the globe. Public research funding programs in Europe, America and Asia represent an important source of content. The selection of work for publication is done under the supervision of an international scientific advisory board that brings together some of the most influential technology leaders and scholars of our time. The target audience primarily consists of industry professionals and academic researchers working at the forefront of their fields, but the book series will also be of interest to advanced level and PhD students.

More information about this series at http://www.springer.com/series/11573

Gereon Meyer · Sven Beiker
Editors

Road Vehicle Automation 6

Springer

Editors
Gereon Meyer
VDI/VDE Innovation + Technik GmbH
Berlin, Germany

Sven Beiker
Stanford University
Palo Alto, CA, USA

ISSN 2196-5544 ISSN 2196-5552 (electronic)
Lecture Notes in Mobility
ISBN 978-3-030-22932-0 ISBN 978-3-030-22933-7 (eBook)
https://doi.org/10.1007/978-3-030-22933-7

This Springer imprint is published by the registered company Springer Nature Switzerland AG
The registered company address is: Gewerbestrasse 11, 6330 Cham, Switzerland

Preface

The great promise that came with the invention of the automobile more than a hundred years ago was that it would do away with the limitations of former mobility solutions like railways or horse carts and to allow its driver and passengers to go faster and reach any place at any time they wished. Automation of vehicles is now set to fully deliver on this promise by a vision of unprecedented road safety in combination with better efficiency, convenience, and inclusiveness. However, these advancements are not without new limitations: A car in a conditional automated mode according to SAE will call its driver to take over control whenever necessary, while a highly automated, driverless shuttle will simply stop should it miss crucial information for making the right decisions. For the automated car to become a desirable product, such issues have to be overcome. Artificial intelligence and connectivity with sensors in the infrastructure, databases in the cloud and centralized traffic management are now widely anticipated to provide solutions for handling critical situations. It is worth noting, though, that all these are highly systemic solutions requiring digital infrastructures and networks that even in the long term will not be available everywhere at any time. Therefore, it may be necessary to rethink the concept of automation and to create seamless transitions between different modes and levels of automation—a challenge that will require technologists, planners, and regulators to cooperate.

The journey of developing the technical, legal, economic, social, and human-centered foundations of road vehicle automation has only just begun. The books on Road Vehicle Automation, which we have been editing since 2013, are covering in a comprehensive way the progress that has been made and the new questions that have arisen. Some authors have been contributing every year, allowing the readers to closely follow the innovation process. Other authors are returning after a few years to share the results of the work they outlined previously. Ourselves, we always aim to point to some important issues in this editorial. While in the first volume, we acknowledged SAE International's definition of automation levels, now, we would like to emphasize the concept of the operational design domain (ODD) that SAE International has added in a recent update to describe the boundaries of the system functionality at a certain level of automation. It is extremely helpful to describe the conditions for the seamless transition between automation levels claimed above, namely as an overlap of complementing ODDs.

The book at hand is the sixth edition of the Road Vehicle Automation books that Springer has published as part of the Lecture Notes in Mobility series. The content covers public sector activities, human factors aspects, vehicle systems, and other technology developments, as well as transportation infrastructure planning. All chapters are based on oral and poster presentations at the Automated Vehicles Symposium (AVS) 2018 in San Francisco, California (USA). At this point, we would like to express our deep gratitude to the organizers of AVS, Jane Lappin and Steve Shladover, for all their outstanding support of this publication over the recent years. Special thanks also go to the Applied Sciences editorial team at Springer, to the Association of Automated Vehicle Systems International (AUVSI) and to colleagues at VDI/VDE-IT for helping to make this publication available just in time for this year's AVS conference. As a side remark, we would like to mention that a successful German-US networking event on future mobility jointly moderated by us has been organized by the German American Chamber of Commerce in San Francisco on the release of all Road Vehicle Automation books since 2017.

Last but not least, we acknowledge the time and effort of all authors who wrote and updated their manuscripts for the chapters of this book. The tremendous circulation of the e-book versions in more than 300 libraries on the world, hundreds of thousands of chapter downloads, and hundreds of citations of the content prove the excellence of their work.

May 2019

Gereon Meyer
Sven Beiker

Contents

Introduction: The Automated Vehicles Symposium 2018

Steven E. Shladover[1(✉)] and Jane Lappin[2]

[1] University of California PATH Program,
1357 South 46th Street, Building 452, Richmond, CA 94804, USA
steve@path.berkeley.edu
[2] Toyota Research Institute, 1 Kendall Square, Cambridge, MA 02142, USA
Jane.Lappin@tri.global

Abstract. The 2018 Automated Vehicles Symposium built on the successes of the predecessor meetings, with an even larger and more diverse roster of participants and a broader selection of breakout sessions. The plenary and poster presentations and breakout discussions continued to provide the meeting participants with the most up-to-date and authoritative information about the current international state of development and deployment of road vehicle automation systems, making this the essential meeting for industry, government and research practitioners in the field.

Keywords: Road vehicle automation · Road transport automation · Automated vehicles · Autonomous vehicles · Self-driving vehicles

1 Overview

The 2018 Automated Vehicles Symposium was organized and produced through a partnership between the National Academies of Science, Engineering and Medicine (NASEM) Transportation Research Board (TRB) and the Association for Unmanned Vehicle Systems International (AUVSI), continuing the pattern established in the four preceding years. This meeting was organized to serve their constituencies' interests in understanding the impacts, benefits, challenges and risks associated with increasingly automated road vehicles and the environments in which they operate. It brought together key government, industry and academic experts from around the world with the goal of identifying opportunities and challenges and advancing automated vehicle (AV) and highly automated driving (HAD) research across a range of disciplines.

The symposium took place over five days, 8–12 July, 2018 with four days of core activities and ancillary sessions on the first day. The morning plenary sessions included presentations from the public sector, automakers and suppliers and research institutes and the afternoons were devoted to thirty-five breakout sessions for deeper investigation and discussion of selected topics. Receptions and poster sessions followed the close of the breakout sessions on Monday, Tuesday and Wednesday afternoons.

The breakout sessions were each organized by committees of volunteers to address a wide range of topics. Four of the breakout sessions spanned two afternoons of the Symposium, providing more time for exploration in greater depth and breadth:

G. Meyer and S. Beiker (Eds.): AUVSI 2019, LNMOB, pp. 1–8, 2019.
https://doi.org/10.1007/978-3-030-22933-7_1

- Trucking Automation: Deployment Challenges and Opportunities
- Enabling Technologies
- Safety Assurance of Automated Driving
- Energy and Emissions Implications of Connected and Automated Vehicles

The other breakout sessions covered a single afternoon each. These are clustered based on their general subject matter as follows:

Transportation User Issues

- The Role of Human Factors in the Design of AV External Communications
- Training Needs for Automated Driving
- AVs and Vulnerable Road Users: Healthy, Safe and Equitable Future
- AVs: Driving Employment for People with Disabilities
- User Needs, Accessible Design and Deployment Challenges to Maximize Benefits
- Evidence-Based Behavioral Studies of Impacts of AV Systems

Transportation/Traffic Operations Issues

- From Automated Vehicles to Automated Transportation Systems
- New Innovations in Intersection Control with Cooperative Automation
- ITS Architecture and Standards Evolution to Integrate Automated Driving
- OEM/DOT Dialogue on Dedicated Lanes, Work Zones and Shared Data
- Road Weather Management: Automation Technology and Adverse Weather

Modal Applications

- Life in the Slow Lane: Automated Low-Speed Shuttles
- Preparing for AVs and Shared Mobility
- Linking Automation, Sharing and Public Transportation

Technology Topics

- Cybersecurity of the AV Ecosystem
- Blockchain in the AV Ecosystem
- AI and Deep Machine Learning Tools and Algorithms for CAVs
- Accelerating AV Market Penetration by Leveraging Infrastructure
- Reading the Road Ahead: Traffic Control Devices and Machine Vision
- Digital Infrastructure: Building a Shared Vision for National AV Readiness

Planning and Impact Assessment

- Building Automation into Urban and Metropolitan Mobility Planning
- Integrating AVs on City Streets: An Interactive Planning Workshop
- "Shark Tank II"

Policy and Regulatory Topics

- Policy Implications of AVs
- Ethical and Social Implications of AVs
- A License to Drive: How to Allow AVs on Public Roads
- Data Exchanges to Enable AV Integration
- An AV Crash Occurs: What Happens Next?

- "Speed Dating" in the Legal Coliseum
- Test Scenarios and Standards for Automation
- Testbed Evolution and Collaboration: A Necessary Path to Roadworthiness

The symposium also involved several related meetings that occurred before and following the main meeting:

- U.S. DOT Listening Session
- National Cooperative Highway Research Program panel 20-102, sponsoring research on impacts of connected vehicles and automated vehicles on state and local transportation agencies
- SAE On-Road Automated Driving (ORAD) Standards Committee meeting
- Meeting of the TRB Committee on Automated Transit Systems (AP040)
- Meeting of the TRB Forum on Preparing for Automated Vehicles and Shared Mobility Systems
- U.S. – Japan – EU Trilateral Working Group on Automation in Road Transportation
- Pegasus Project International Workshop

In keeping with TRB practice, the plenary and breakout sessions were planned and produced by volunteers whose expertise and work informed the content of the sessions. In keeping with AUVSI practice, the production of the symposium was professionally managed by dedicated conference and logistics managers. The AVS18 Executive Committee reflected this mix of the two organizations.

Richard Bishop, AUVSI subject matter expert on automation; Richard Cunard, Senior Program Officer, Traffic and Operations Engineer, TRB; Bob Denaro, ITS Consultant, Chair, TRB Joint Subcommittee on the Challenges and Opportunities for Road Vehicle Automation; Thomas Dingus, Virginia Tech Transportation Institute and AUVSI Board Member, Kevin Dopart, U.S. DOT Intelligent Transportation Systems Joint Program Office, Ginger Goodin, Texas A&M Transportation Institute, Jane Lappin, Toyota Research Institute; Jack Pokrzywa, Director, SAE Global Ground Vehicle Standards; Steven Shladover, University of California PATH Program, Chair, TRB Vehicle-Highway Automation Committee (AHB30); Lindsay Voss, Senior Program Development Manager, AUVSI.

2 Symposium Attendees

About 1650 registrants participated in the symposium, growing by about 150 people over 2017 and continuing the growth experienced over the preceding four years of meetings. Attendees represented a wide range of organizations from government and industry to the academic-, public-, and private-sector research communities. One of the strengths of the meeting was the breadth of interests represented, including industry, public agencies and academic/research organizations. The automobile industry was well-represented with many attendees from Original Equipment Manufacturers (OEMs) and their suppliers.

These participants represented disciplines ranging from engineering to psychology to law. Thirty-two countries (representing the 23% of the meeting participants who come from outside the U.S.) and forty-three U.S. states were represented among the meeting participants. The largest delegation from outside the U.S. came from Japan, with 65 participants, while Germany, Canada, the UK and the Netherlands all had substantial attendance. Consistent with the previous meetings, California, as the host state, had the largest number of attendees from within the U.S., followed by the national capital region (DC, Maryland, and Virginia) and Michigan.

3 Keynote Talks

AVS19 had the benefit of two keynote addresses from the U.S. Department of Transportation. USDOT Secretary Elaine L. Chao presented the first keynote address on Tuesday, July 10. NHTSA Deputy Administrator Heidi King delivered the second keynote address on Thursday July 12.

Secretary Chao opened her address by observing that automated vehicles will have far reaching impacts on transportation. She said that it can save many lives and has the potential to increase access for underserved communities. She affirmed that it is USDOT's regulatory responsibility to understand automated technologies and to help prepare for them.

Secretary Chao enumerated six principles that govern the Department of Transportation's approach to AV technology:

1. The Department's primary priority is safety
2. The Department will be technology-neutral. It will not pick winners and losers among the developers of these technologies.
3. The Department's preference is for regulations that are non-prescriptive, performance-based, and seek to enhance safety whenever possible.
4. The Department will work with states and localities to avoid a patchwork of rules that could inhibit innovation and make it difficult for AVs to cross state lines.
5. The Department will provide stakeholders with guidance, best practices, pilot programs and other assistance to facilitate the safe integration of AV systems into the transportation system.
6. The Department recognizes that highly automated vehicles will have to operate side-by-side with traditional vehicles, in both urban and rural areas.

Secretary Chao highlighted key societal and technical challenges, and shared some of the feedback from an AV policy listening summit that was held at DOT headquarters on March 1, 2018. She said that there was consensus that the Department must lead in facilitating communication with stakeholders and working with the states to develop consistent policies.

Secretary Chao closed her remarks saying that she looks forward to working with the AVS audience in support of the Department's goal to encourage the safe deployment and integration of AV technology, while maintaining maximum individual choice and enhanced safety.

NHTSA Deputy Administrator Heidi King gave the second keynote address on Thursday July 12, opening the final day of AVS18. Ms. King echoed the Secretary's priorities of safety, infrastructure, and preparing for the future, and emphasized

NHTSA's mission to save lives and prevent injuries. In her address, she focused on three key safety-related issues: voluntary disclosure, wireless spectrum, and cyber security. She also informed the group about some of NHTSA's current activities.

Voluntary Disclosure: Ms. King focused on the Department's voluntary guidance: Automated Driving Systems: A Vision for Safety. She said that the 12 safety elements included in the Voluntary Safety Self-Assessment (VSSA) allow the public to see that safety considerations are built into the design and manufacture of driving automation systems. The public wants innovators to assure them that automated driving systems won't crash into things. DOT and NHTSA applaud the companies that have filed their VSSA.

Wireless Spectrum: The Department of Transportation is technology neutral relative to communication protocols, but remains supportive of priority use of the 5.9 GHz band for safety and mobility. The Department is working closely with the FCC and NTIA to explore optimal use of the 5.9 GHz band, and is exploring a variety of technologies, including cellular V2X, or CV2X as well as DSRC.

Cyber security: Ms. King stressed the need to implement best practices and share information on possible threats and vulnerabilities. As an example, she highlighted her experience with the sixth Cyber Storm series, the Department of Homeland Security's biennial exercise. Cyber Storm 6 focused on manufacturing and transportation sectors, and included the Automotive Information Sharing and Analysis Center (Auto-ISAC), other federal agencies and companies. Ms. King said that the lessons learned from that exercise will strengthen the nation's cyber security framework.

NHTSA Activities: Ms. King highlighted a few current actions to support safety innovation.

- NHTSA is seeking public comment on proposals for the establishment of the "Pilot Program for Collaborative Research on Motor Vehicles with High or Full Driving Automation."
- It is seeking comment on existing motor vehicle regulations in "Removing Regulatory Barriers for Automated Driving Systems."
- "Updating the Petition Process for Exemption from Federal Motor Vehicle Safety Standards" is seeking comment on whether updating certain provisions of NHTSA's regulation regarding the processing of petitions for exemption from Federal motor vehicle safety standards (FMVSS) could improve the process for reviewing innovative safety technologies.

Ms. King closed her address by offering a note of caution, advising the audience to recognize the risks together with the benefits of automated driving, to try to make sure that it will contribute to both safety and mobility.

4 Plenary Panel Sessions

Steven Shladover chaired a plenary panel session on safety assurance, validation and certification for automated driving systems, with panelists. Olaf van den Camp, TNO, Phil Koopman, CMU, Tom Dingus, VTTI, and Lutz Eckstein, RWTH Aachen.

Ginger Goodin chaired a panel on regulatory issues for automated driving with panelists Cathie Curtis, AAMVA, Jill Ingrassia, AAA, Edwin Nas, Netherlands Government, David Strickland, Self-Driving Coalition for Safer Streets, and Nat Beuse, NHTSA.

Bob Denaro chaired a panel session on investor perspectives on automated driving with panelists Jim Adler, TRI, Jim DiSanto, Motus Ventures, Amy Gu, Hemi Ventures, Ulrich Quay, BMW Ventures, and Olaf Sakkers, Maniv Mobility.

5 Plenary Presentations

Recent Developments in Vehicle Automation Technology:

- Wherever, Whenever, Whatever the Weather: On-Road and Off-Road Autonomy – Paul Newman, Oxbotica
- Development of Unsupervised Self-Driving Using Deep Learning Neural Networks and Proving It Is Sufficiently Safe – Jonas Ekmark, Zenuity
- The Cambrian Explosion of AV Sensors – Robert Seidl, Motus Ventures
- PEGASUS Program on Safety Assurance – Walther Wachenfeld, Continental

Development and Assessment of Vehicle Automation Concepts and Systems:

- L3 Pilot Project – Adrian Zlocki, RWTH Aachen
- Applying a Hybrid Network Approach to Deployment of Self Driving Mobility Services – Nadeem Sheikh, Lyft
- GATEway: Exploring How People Respond to, Engage with and Accept CAVs in a Challenging Urban Environment – Richard Cuerden, TRL Academy
- Reshaping Urban Mobility with Autonomous Vehicles: Lessons from the City of Boston – John Moavenzadeh, World Economic Forum
- Real Commercial Platooning – Real Safety and Savings – Josh Switkes, Peloton
- Embark's Approach to Operational Safety for Automated Truck Development – Alex Rodrigues, Embark

Identifying and Addressing Key Non-Technological Research Questions:

- Bridging the Automated Vehicle Gap: Consumer Trust, Technology and Liability – Kristin Kolodge, J.D. Power and Tina Georgieva, Miller, Canfield, Paddock and Stone, PLC
- Transition from Collision Avoidance to Automated Vehicle Systems – Ensar Becic, NTSB
- Advancing the AV Opportunity: NHTSA, NTSB and Industry Lessons Learned – Mark Rosekind, Zoox
- Preparing U.S. Workers and Employers for an Autonomous Vehicle Future – Erica Groshen, Cornell University
- Designing Automated Vehicles Around Human Values – Chris Gerdes, Stanford University
- Designing Cooperative Interaction of Automated Vehicles with Other Road Users: Overview of the EU Project interACT – Natasha Merat, University of Leeds

Public Sector Activities on Road Vehicle Automation:

- SIP-adus: Field Operational Tests and Regulatory Issues – Hajime Amano, ITS Japan
- Automated Mobility: The EU Strategy for Mobility of the Future – Clara de la Torre, European Commission
- The Future of Mobility in the United Kingdom - Iain Forbes, UK Department for Transport
- Dubai Self-Driving Transport Vision and World Challenge – Ismail Zohdy, Dubai Road and Transport Authority
- U.S. Department of Transportation Automated Vehicle Research Activities – Finch Fulton and Kevin Dopart, U.S. DOT.

6 Breakout Sessions

The breakout sessions provided opportunities for more in-depth consideration of specific topic areas among groups of people with focused interests in those areas. With smaller groups, they could be more interactive than the large plenary sessions, providing ample opportunities for questions and answers and debates. The primary findings from each afternoon's breakout discussions were reported back to the plenary the following morning.

The breakout sessions covered a wide range of specialized topics relevant to automated driving to match the interests of different groups of meeting participants. The expansion of the breakout session times from two afternoons in previous years to three afternoons this year provided the opportunity to accommodate a wider range of topics, as already listed above in the Sect. 1 Overview.

The report-outs to the plenary session revealed some of the important lessons learned from the breakout discussions that have broad significance across the field:

- the dominance of uncertainty about many aspects of AV system capabilities, timing and impacts;
- the need to define basic performance requirements and standards for various aspects of AV systems, coupled with the great difficulty of doing so at this stage in the development of the AV technology and operational concepts;
- the need to educate and manage the expectations of the general public, media and decision makers about AV systems;
- the need for more research and real-world data regarding many aspects of AV capabilities and operations.

7 General Cross-Cutting Observations

As the field of road vehicle automation has advanced and the level of knowledge of the issues has grown over the past several years, the areas of emphasis within the Automated Vehicles Symposium have shifted. In this most recent meeting, several general observations are worth noting:

- The meeting attendance and participation demonstrated the breadth of interest in the subject of road vehicle automation across different sectors of society (private industry, governments and research organizations) and across countries and continents.
- The presentations and discussions revealed a growing recognition of how long a time it is likely to take for the more advanced automation capabilities to be deployed based on unresolved technological challenges. This led to more emphasis than in previous years on some of the intermediate levels of automation that will have to be shown to work effectively before the more advanced systems will become practicable and acceptable to the broader stakeholder community.
- There was also growing recognition of the central importance of the operational design domains of the automation systems and how the initial implementations of the more highly automated systems will have to start with greatly simplified operational design domains. As the technology improves, the constraints on the operational design domains will gradually be relieved and the systems will be able to provide transportation services under a wider range of conditions.
- In parallel with the preceding points, there was also growing attention to the need to communicate clearly and accurately with the general public, the media and thought leaders about the capabilities and limitations of the automation technology so that they will have realistic expectations. If the stakeholder expectations are not consistent with the technological realities, there could be problems with unsafe misuse of systems and of public backlash when unrealistic expectations are not met by the systems that are available for public use.
- The importance of safety assurance, verification and validation challenges was more widely recognized, and these topics received more consideration on both the technical and regulatory sides. Technically feasible and publicly credible approaches to safety assurance will be essential prerequisites to public implementation of highly automated driving systems.

Part I Public Sector Activities

SIP-adus: Field Operational Tests and Regulatory Issues

Hajime Amano(✉) and Takahiko Uchimura

ITS Japan, 2-6-8 Shibakouen, Minatoku, Tokyo 105-0011, Japan
{h-amano, t-uchimura}@its-jp.org

Abstract. Large scale field operational tests were conducted under the Japanese national project on Connected and Automated Driving, SIP-adus (Strategic Innovation Promotion Program – Automated Driving for Universal Services). Integration of dynamic traffic data through radio communication with high-definition map data was one of the important focuses of the tests. The first phase of SIP-adus, presented at AVS 2018, was completed in March 2019 and the second phase was initiated with additional set of objectives; deployment of passenger and freight services, development of Dynamic Map data management framework and development of technologies for safety validation and certification. Revision of regulatory framework and deployment scenario for socioeconomic benefits are also vigorously investigated.

Keywords: Automated driving · Dynamic map · Connected services · Field operational test · Low speed automated shuttle · Truck platooning · Inclusive society

1 The First and the Second Phase of SIP-adus

SIP-adus is a Japanese connected and automated driving project. The technology elements of SIP-adus are shown in Fig. 1.

1.1 Field Operational Tests

1.1.1 Overview of the Field Operational Tests

In 2018, as the final year of the first phase of the 5-year project, large scale Field Operational Tests were conducted [1]. At the same time, the second phase of 5-year SIP-adus started in 2018. Technologies developed in sub-projects were integrated into 5 themes of the Field Operational Tests:

- Dynamic Map
- Human Machine Interface (HMI)
- Cyber Security
- Pedestrian Accident Reduction
- Next Generation Transport

International cooperation was also an important aspect of the field operational tests. A number of international participants (OEMs, suppliers and research institutes)

G. Meyer and S. Beiker (Eds.): AUVSI 2019, LNMOB, pp. 11–17, 2019.
https://doi.org/10.1007/978-3-030-22933-7_2

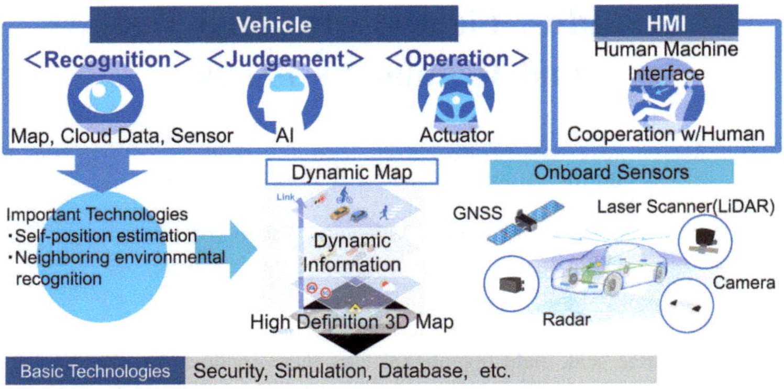

Fig. 1. Overview of the SIP-adus research and development

signed up. Dynamic Map data and on-board equipment to receive connected services were distributed to the participants. Through the tests on the common grounds, concrete evidences were acquired and the participants actively joined in-depth discussions on the research topics to identify shared challenges and direction to overcome them.

Three categories of the test sites were selected in the Tokyo Metropolitan area;

- 300 km stretch of Expressways
- Arterial roads in Tokyo waterfront area
- Dedicated test facilities separated from the general traffic.

1.1.2 Integration of High Definition 3D Map with Dynamic Data

Automated vehicles make decisions comparing high definition 3D map with the data acquired by on-board sensors as shown in Fig. 2. High definition 3D map database for the entire road environment of the test sites were developed and delivered to the participants. Participants drove the highway and evaluated the accuracy of the 3D map. They also actively engaged in the discussion on data elements, format and update frequency to build consensus. Dynamic data provided through wireless communication were also added to enhance the 3D map into the Dynamic Map. In Japan, V to I services have long been in operation. Those existing connected services were utilized for evaluation of functionality in the context of automated driving.

Using the same spectrum as Electronic Toll Collection, about 3,600 DSRC roadside beacons are operating. About 2.4 million human driven cars and trucks are equipped with V to I capable on-board equipment. On arterial roads, Traffic Signal Prediction System is operating at more than 7,000 intersections. Phase and timing of traffic signals are broadcast at those intersections. On-board system judges safe and most efficient speed and acceleration or deceleration timing. The collision warning system alerts the driver about the presence of other road users. These connected services were integrated as a part of the Field Operational Tests on the arterial roads as shown in Fig. 3.

Those connected services are using several different wireless communication technologies, which were selected at the time of deployment for good technical reasons, more than 20 years ago, in some cases. Field Operational Tests under SIP-adus

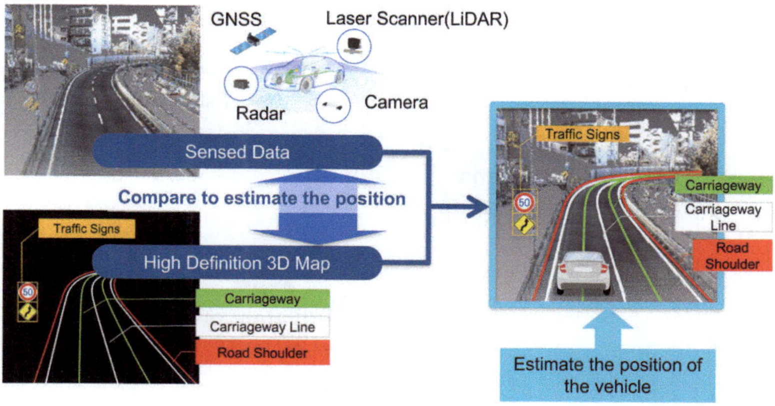

Fig. 2. Vehicle position detection using dynamic map

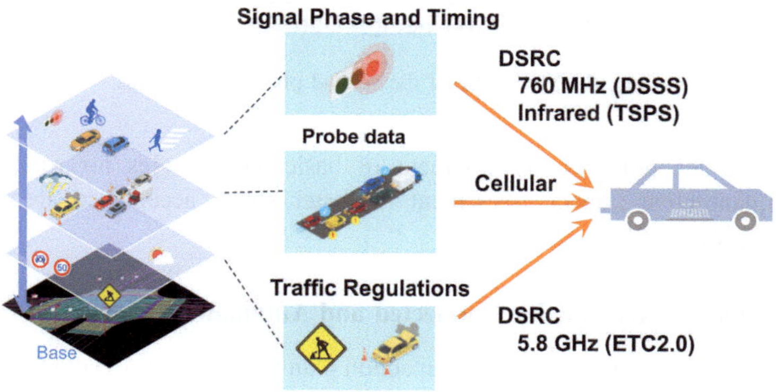

Fig. 3. Dynamic map evaluation with connected features

were not intended to evaluate wireless communication technologies. It was the intention to identify potential benefits and challenges in integrating dynamic data into automated driving systems. Communication devices to interface the participant's vehicles with those data were provided by the government. The Field Operational Tests concluded the 1st phase of SIP-adus. Comprehensive reports were published as 'SIP-adus: Project Reports, 2014–2018 - Automated Driving for Universal Services' in March 2019.

1.2 The Second Phase of SIP-adus

The 2nd phase of SIP-adus started with different focuses as shown in Fig. 4 [2, 3].

Operational Design Domain is extended from highways to arterial roads. More resources are put on mobility services. Societal benefits of automated driving are vigorously investigated. Starting with deployment for the Tokyo Olympic and Paralympic Games, local government and private sector are expected to follow. Validation

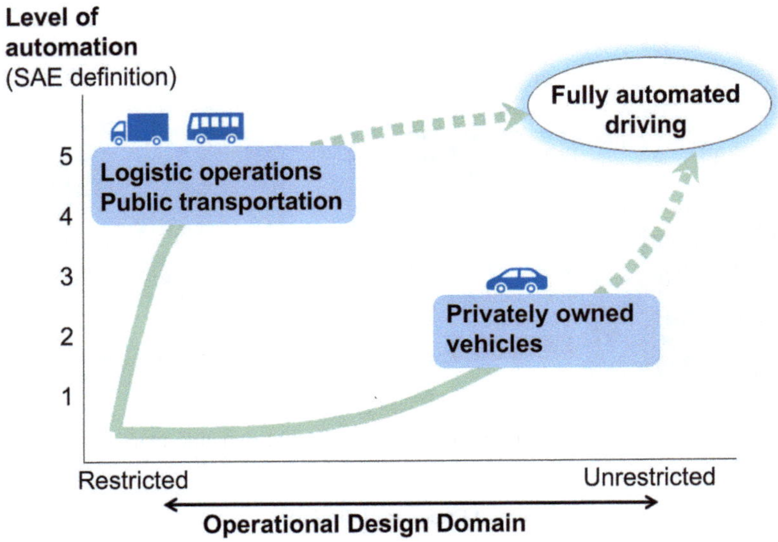

Fig. 4. The scope of the second phase of SIP-adus

of integrated technologies, development of basic technologies for roadworthiness testing, quantitative impact assessment to foster social acceptance are the major research topics.

1.3 Regulatory Reforms for Connected and Automated Driving

The Japanese government authorized an action plan of regulatory development entitled 'The charter for improvement of legal system and driving environment for automated driving systems'.

The Japanese government takes a holistic approach to ensure safety as shown in Fig. 5. Vehicle technologies are rapidly improving. Road infrastructure needs to be improved to meet the new set of requirements. Restriction of operational domain is considered. Intervention by the human driver is assumed to ensure sufficient level of safety. The required level of safety collectively will be achieved with changing composition of each element over time [4].

Actions are being taken so that regulatory framework becomes ready for deployment of SAE Level 3 automated driving by 2020. Vehicle safety regulations, conformance testing and type approval procedures will be developed. As the first step, safety guidelines were released in September 2018. Road traffic rules also need to be revised for automated driving. Rules for automated driving vehicle with remote monitoring and for truck platooning are within the scope of consideration. Automobile liability insurance will be applied for immediate relief of victims. Insurance companies will reimburse from the responsible entities. Criminal responsibilities and installation of Event Data Recorders will be investigated. Regulations for public transportation and freight operators are reviewed [5].

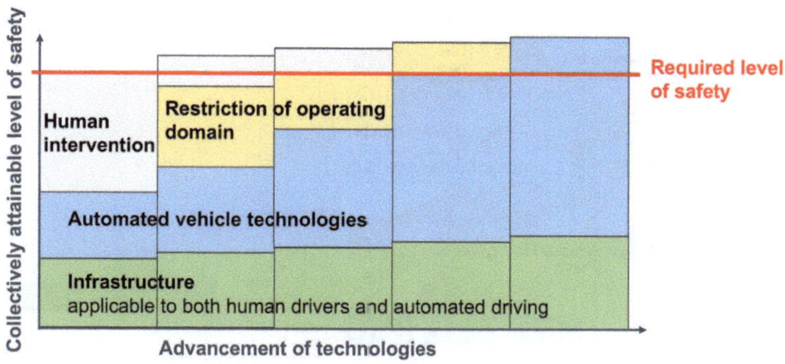

Fig. 5. Holistic approach for safety

2 Societal Benefits of Connected and Automated Driving

2.1 Societal Challenges in Japan

The most serious challenge for the Japanese society is ageing and declining population. According to the report compiled by the government, by the year 2050, 19% of now inhabited area will be totally deserted where nobody is living. Another 44% will lose more than half of the current population [6].

ITS Japan, a non-profit organization coordinating collaboration between the private sector and the public sector, compiled a vision of integrated mobility services to tackle societal challenges, through discussions across the industrial sectors and across the academic disciplines. The vision describes societal benefits expected in Japan by deployment of automated driving as shown in Fig. 6. In rural areas, where most serious population decline is projected, small villages are connected to a basic social service hub with transportation and information network. Middle size cities are integrated to have combined population of at least 300,000, necessary to maintain high level education, medical care and employment opportunities. They are connected to each other within one hour of travel.

2.2 Low Speed Automated Shuttle Service

Ministry of Land Infrastructure, Transport and Tourism is conducting field operational tests in rural areas as shown in Fig. 7. Automated shuttles for elderly people who no longer drive cars also deliver goods to and from the social service hub [7].

Automated shuttle services in rural areas will provide elderly people with mobility to support their daily lives. It will also provide small scale farmers and traditional craftsmen and women with opportunities to reach consumers living in big cities and keen on specialized products. New businesses without restriction on their office location are setting up satellite offices in rural areas for better environment for creativity and family life. Migration of people will help rural communities to survive. Innovation in technology will contribute to our society more effectively when it is integrated with societal factors.

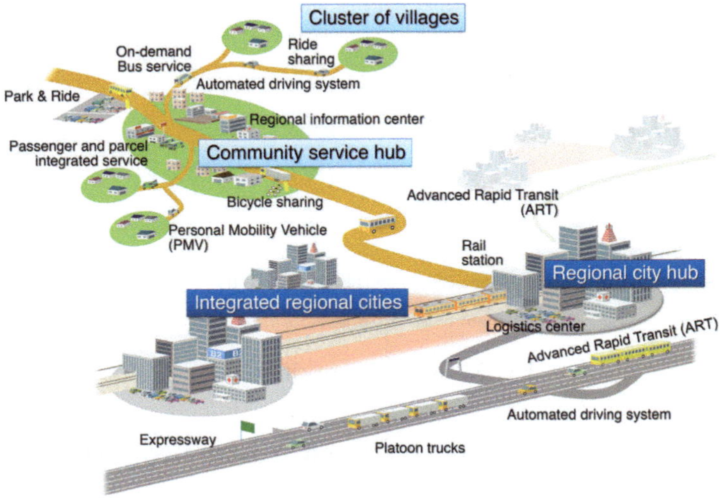

Fig. 6. A vision of integrated mobility services compiled by ITS Japan

Fig. 7. Low speed automated shuttle service in rural areas

2.3 Automated Platoon of Trucks

Fully automated truck platoon system was developed and demonstrated in 2012. CACC truck platoons in mixed traffic on highway are being tested for deployment [8].

Today, the entire chain of manufacturing processes or consumer goods delivery is synchronized, leveling workload and flow of materials. It is called the Lean production system. Other keywords followed, such as Agile system and Resilient system. So, simply deploying platoon of large trucks carrying large lot of goods could have negative impacts on the entire operation. Truck platoon is only a part of the picture. It is essentially important to create another innovative operation concept to fully exploit benefits from automated driving technologies as shown in Fig. 8.

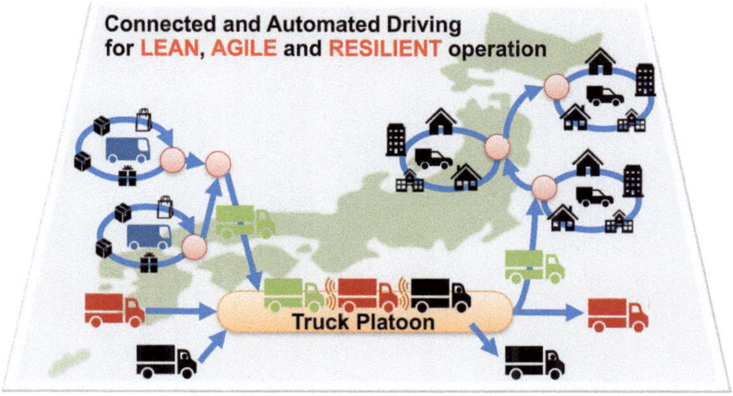

Fig. 8. Effective integration of automated platoon of trucks in supply chains

3 Conclusion

At the beginning of SIP-adus in 2014, the objectives were stated; to create inclusive society, where automated driving technologies integrated with social innovations provide everyone with mobility to fully exercise his or her capacity, enabling sustainable development of the society. This is exactly the direction we pursue in the second phase of SIP-adus.

References

1. Minakata, M.: SIP-adus field operational test. In: SIP-adus Workshop, Tokyo (2018)
2. Koga, Y.: SIP-adus project, national research and development initiative on automated driving in Japan: outcomes and next actions. Transportation Research Board Annual Meeting 2019, Washington, D.C. (2019)
3. Kuzumaki, S.: SIP automated driving systems, keynote. In: SIP-adus Workshop, Tokyo (2018)
4. Hachiyama, K.: ITS development policies in Japan. In: SIP-adus Workshop, Tokyo (2018)
5. Sugi, T.: Current trend and NPA initiatives regarding automated driving. Transportation Research Board Annual Meeting 2019, Washington, D.C. (2019)
6. Amano, H.: Technological evolution leading to societal innovation. ITS Summit Singapore 2018, Singapore (2018)
7. Itsubo, S.: First-mile/last-mile automated shuttle research in Japan. Transportation Research Board Annual Meeting 2019, Washington, D.C. (2019)
8. Kakimi, N.: Truck platooning project in Japan. Transportation Research Board Annual Meeting 2019, Washington, D.C. (2019)

Part II Human Factors and Challenges

How Do We Study Pedestrian Interaction with Automated Vehicles? Preliminary Findings from the European interACT Project

Natasha Merat[1(✉)], Yee Mun Lee[1], Gustav Markkula[1], Jim Uttley[1], Fanta Camara[1,2], Charles Fox[2], André Dietrich[3], Florian Weber[4], and Anna Schieben[5]

[1] Institute for Transport Studies, University of Leeds, Leeds LS2 9JT, UK
n.merat@its.leeds.ac.uk, {y.m.lee, g.markkula, tsfc}@leeds.ac.uk, j.uttley@sheffield.ac.uk
[2] School of Computer Science, University of Lincoln, Lincoln, LN6 7TS, UK
chfox@lincoln.ac.uk
[3] Department of Engineering, Chair of Ergonomics, Technical University of Munich, Boltzmannstasse 15, 85748 Garching, Munich, Germany
andre.dietrich@tum.de
[4] Innovations BMW Group, Munich, Germany
Florian.WW.Weber@bmw.de
[5] DLR e.V. - German Aerospace Centre, Lilienthalplatz 7, 38108 Braunschweig, Germany
anna.schieben@dlr.de

Abstract. This paper provides an overview of a set of behavioural studies, conducted as part of the European project interACT, to understand road user behaviour in current urban settings. The paper reports on a number of methodologies used to understand how humans currently interact in urban traffic, in order to establish what information would be useful for the design of future AVs, when interacting with other road users, especially pedestrians. In addition to summarising the results from a number of observation studies, we report on preliminary results from Virtual Reality studies, investigating if, in the absence of a human vehicle controller, externally presented interfaces can be used for communication between AVs and pedestrians. Finally, an overview of the mathematical and computational modelling techniques used to understand how AV and pedestrian behaviour can be both cooperative, and effective is provided. The hope is that future AVs can be designed with an understanding of how humans cooperate and communicate in mixed traffic, promoting good traffic flow, user acceptance and user trust.

1 Introduction

An inherent challenge in mixed traffic environments of the future is that manually driven and automated vehicles (AVs) will need to interact with non-automated road users, such as pedestrians and cyclists. This interaction may occur in ambiguous

G. Meyer and S. Beiker (Eds.): AUVSI 2019, LNMOB, pp. 21–33, 2019.
https://doi.org/10.1007/978-3-030-22933-7_3

scenarios, where the rules of the road may not be clear, either due to lack of clear environmental/infrastructural advice, and/or as a result of local or national cultural and behavioural "norms". Here, there is typically a need for cooperation, and constructive communication and interaction between these different actors, so that they may reach an agreement regarding safe future motion plans, especially if they share the same road space. Such interactions are currently quite frequent in an urban environment, and with the introduction of "driverless" AVs, it is important for all road users to have a good understanding of the intentions of these vehicles, especially in the absence of an accountable human operator. Therefore, there is a need to understand how the right cooperation strategy between all road users can be developed, to ensure successful deployment and acceptance of AVs by all road users, and promote smooth and cooperative flow of traffic.

The vision of the interACT project[1] (https://interact-roadautomation.eu), funded by the European Commission, is to develop novel, holistic interaction concepts for AVs, that will enable the future integration of these vehicles in mixed traffic environments, in a safe and intuitive manner. Currently, as road users, humans use multiple means of implicit cues, such as approach speed, and explicit communication, such as eye contact and gestures, as well as vehicle signals, to anticipate the intention of the other traffic participants on the road. Although the exact means of communication can differ across different regions and cultures, these acts allow effective coordination of future motion plans between different road users. However, currently, AVs are thoroughly lacking such coordination capabilities, and their interaction with other road users is often limited to, and mostly dominated by, the rational principle of collision avoidance. Therefore, to safely integrate AVs in complex, mixed traffic environments in the future, we must ensure that the AV can interact with other road users in an intuitive, expectation-conforming manner. This will allow other road users, as well as those on-board the AVs (who may still be required to resume control in case of emergencies), to correctly interpret the intentions of the AV, and coordinate their planned actions accordingly. Results from a previous study, conducted during the CityMobil2 project [1], showed that when interviewed after interacting with low speed AVs (which operated in a shared space setting, and without a driver) pedestrians and cyclists highlighted the importance of *some* kind of external communication messages from these AVs, to compensate for the absence of an accountable operator[2]. A message that acknowledged they had been detected by the AV was rated highest by this group of 664 respondents, interviewed across Greece, France and Switzerland. As a follow-on to some of the human factors questions addressed in the Citymobil2 project, interACT is conducting further work in this context, to enhance knowledge in the field, and improve the interaction of AVs with both the on-board user and pedestrians by:

[1] EU H2020 interACT: Designing cooperative interaction of automated vehicles with other road users in mixed traffic environments under grant agreement No 723395.

[2] For safety reasons, an "operator" was present in these vehicles. This was a trained individual, who was responsible for managing the vehicle's manoeuvres during difficult circumstances, such as intervening when movements were required to avoid obstacles which appeared unexpectedly in the vehicle's path (such as parked cars).

- Developing psychological models of interaction as the basis for the development of a "*Cooperation and Communication Planning Unit*", a central software unit for the integrated planning of intuitive AV interaction, based on the AV behaviour, and explicit communications with its on-board user and other traffic participants;
- Enhancing methodologies for intention recognition and behaviour prediction of other road users, to allow shared situation awareness, and coordinated and safe vehicle behaviour planning;
- Establishing a safety layer for all situations, in which interaction is not possible/not safe enough, or in case of interaction failures, i.e. due to misinterpretations;
- Developing novel fail-safe trajectory planning methods, with a special focus on complex mixed traffic scenarios;
- Establishing new evaluation methods for studying interaction of road users with AVs, and ensuring user acceptance.

Efforts are currently undergoing by project partners in Germany, United Kingdom, Italy and Greece, to achieve the above goals. This chapter reports on the efforts achieved in the first year of the project, which has included: (i) extensive observation and interview studies conducted in current urban environments, noting the types of interactions and communications taking place between pedestrians and drivers; focusing particularly on low speed environments and un-signalised junctions; (ii) Lidar and video-based analysis to obtain kinematic data regarding road users' interactions in a complex setting; (iii) Human-in-the-loop virtual reality studies, to understand pedestrians' crossing behaviour in response to vehicles travelling at various speeds, investigating whether different types, positions and colours of externally presented messages from AVs affect crossing behaviour; and (iv) mathematical modelling techniques, used to inform AV developers of the types of interactions expected by other road users, and how this can be managed by the AV, to create better traffic flow, and a fairer, yet more cooperative, relationship between different road users sharing the same road space.

The next sections provide a short overview of each of the above investigations, summarising our current understanding of the state of the art, and briefly comparing these to related studies in this context.

2 Human Interactions and Negotiations in Current Urban Settings

2.1 Pedestrian-Driver Interaction at Un-Signalised Junctions

To understand how drivers and pedestrians currently interact with each other at un-signalised junctions, where negotiations are necessary in the absence of clear infrastructure-based guidelines, such as traffic lights and zebra crossings, we started our investigations by observing current behaviour in urban settings across three European cities. A series of on-road observations were conducted in: Leeds, UK; Athens, Greece;

and Munich, Germany; which were also accompanied by birds-eye view video recordings of the junctions[3] (see Fig. 1).

Following a project workshop amongst the partners, effort was made to find a similar setting across the three cities, although practicalities regarding ease of data collection, erection of video cameras, and geographical differences, created some challenges regarding an exact match. The main criterion here was that observations should be based at un-signalised locations, encouraging "jay walking", in order to assess any negotiation tactics used, in the absence of formal traffic rules. Extensive effort was then invested by partners to create an easy to administer, HTML-based observation app see [2, 4], which was comprehensively piloted before data collection, to ensure researcher familiarity. Two observers were then positioned at designated locations in each city, and recorded any observable behaviour by the pedestrians, drivers and their vehicles, using the app. Communication between the observers took place throughout data collection, and the type of data recorded included: body signals from the pedestrians and drivers (hand/looking behaviour), observable messages from the vehicle (such as flashing lights or honking horns), and any "negotiation tactics" by either actor, with regards to the crossing manoeuvre, such as stopping, decelerating, or crossing the road. The app also allowed recording of road user demographics (gender and age category), road details (exact location) and weather. For a more comprehensive overview of the observation protocol, see [2, 5] for more details).

Data from 989 pedestrian interactions were collected by these observations. Overall, results from these studies showed quite similar behaviour by all road users, regardless of country studied. An interesting observation, also confirmed by the work of others in this context [7, 9, 10], was the distinct lack of *explicitly observable* gestures by the pedestrians and drivers, with less than 4% of pedestrians and 3% of drivers using hand or head gestures during the negotiations. Honking and flashing lights were only seen for 1% of the interactions. Instead, results suggest that pedestrians may use the vehicle's behaviour to determine their crossing decision, crossing when they ascertained yielding by the vehicle.

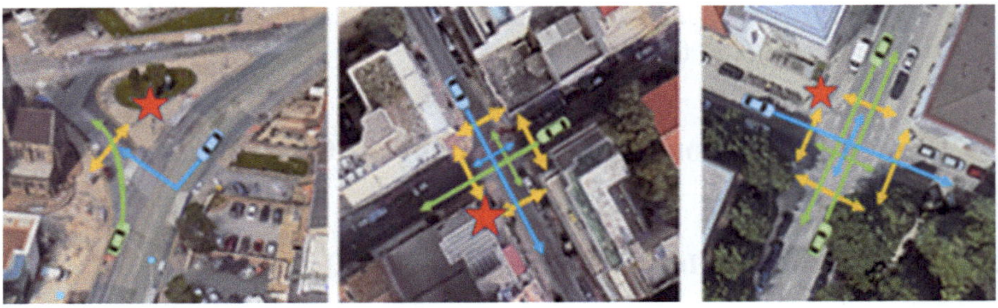

Fig. 1. An aerial view of the intersections used at Leeds (left), Athens (middle) and Munich (right). Yellow arrows represent the location and direction of pedestrians' crossings. The blue and green lines represent the direction of travel for vehicles. The red stars represent the location of a group of two observers who used the mobile app to record observed behaviour (for further details see [2]).

[3] Ethical approval was granted by the University of Leeds ethics committee (Ref: LTTRAN-097).

Indeed, a follow on questionnaire study, administered to a subset of the pedestrians (~20%), after they crossed the junction, confirmed this prediction [2]. An interesting observation across all sites was that, overall, only 72% of pedestrians looked towards the vehicle as they crossed the road. Therefore, this could provide an interesting challenge for the AV, if it is to establish whether pedestrians have identified its presence, before approaching a shared location, where there is no obvious clue from their body language.

2.2 Results from Video- and Lidar-Based Data Analysis

In addition to the observation studies outlined above, video recordings of the interactions were conducted, by placing cameras in an elevated position, overlooking the junctions, as shown in Fig. 1. Computer vision was used for developing detection, classification and tracking algorithms, combined with camera calibration and homography, to extract kinematic data from observed traffic participants (see also Sect. 4.2). A ground based LiDAR was used to record the positions of traffic participants over time, reducing the use of kinematic movements from videos, and removing any personal data. However, challenges existed for use of this ground based LiDAR, due to occasional obstructions. Therefore, the link between these recordings and manual observations were key, to provide a more holistic overview of the interactions.

Analyses from the LiDAR and video data are currently ongoing, although preliminary results suggest that there was a velocity threshold for interactions, where drivers mostly provided pedestrians with a right of way, when their travelling speed was already well below the allowed speed limit. This, more cooperative, behaviour from drivers was mostly observed in congested traffic, and during tailbacks at signalised intersections, because drivers had already reduced their travelling speed, and it was therefore easier to offer a clear path for the jaywalking pedestrians. As discussed further below in Sect. 4.2, such an interaction will be quite problematic for current AVs, as an approach from jaywalking pedestrians will likely result in a yielding action by the AV, to avoid collision, which will likely cause more erratic flow of traffic, especially for other, human-controlled, non-automated, vehicles.

3 Using Virtual Reality to Study Human Interaction with Future AVs

Recently, a number of vehicle manufacturers, keen to deploy driverless "robotaxi" – style AVs without a responsible human controller, have begun discussing the benefit of implementing some type of external message on an AV, which will inform pedestrians about its behaviour – replacing any human-based communication [3]. Our human factors work from the CityMobil2 project provides rather mixed results about the best type of message used in this context. This study showed that, although pedestrians in Greece, Switzerland and France were all keen to receive *some* sort of information from the AV, preference for visual versus auditory messages was rather mixed across different groups. The type of message preferred was also linked to the behaviour depicted by the AV, and varied due to cultural norms, as well as the different infrastructures available to the AV [8].

Overall, however, respondents from this project preferred the use of conventional signals (lights and beeps) to text and spoken words, and wished to receive either visual or auditory signals that would announce information about whether or not the vehicle was turning/yielding/beginning to move. Other work in this area has begun to investigate the matter further, testing a variety of driving conditions, to establish the efficacy of such external messages e.g. [6]. In addition to the above examples, studies have investigated the value of messages that are used to express: whether or not it is safe for the pedestrian to cross, whether AVs that look like a conventional vehicle should signify their automation status, and whether particular types, colours, and locations of lighting are better than others [9–11]. Results have been mixed, with some showing major changes in crossing behaviour, such that pedestrians' receptivity towards AVs significantly increased with the presence of external HMIs [12], and others, for example [10], suggesting that pedestrians rely on the behaviour of the vehicles rather than the information on the external HMI.

In the absence of easily accessible (fully) driverless vehicles, which can be used to portray different types of external interfaces for communication with pedestrians, novel tools such as Wizard-of-Oz techniques [13], human-in-the-loop pedestrian simulators [13], and immersive Virtual Reality (VR) Head-Mounted Displays [14] are used to provide a suitable alternative for cost-effective, controlled and repeatable research studies in this context. In the interACT project, VR has been very effective for such research, evaluating and improving potential interaction strategies between humans and future AVs. Here, design-focused workshops with expert and naïve participants have been used for visualisation of potential solutions, with relatively minimal effort spent on defining and refining external HMIs (eHMIs), before deploying them for actual user studies on prototype vehicles [15].

VR offers the opportunity to study human-AV interaction for assessing the speed and quality of comprehension of AV behaviour, or for assessing the traffic participants' behaviour or emotions in response to the AV. For example, Head- Mounted Displays (HMDs) have been used in the project to assess pedestrians' actual crossing behaviour in VR, in response to vehicles with different kinematic features [2]. This type of manipulation is useful for evaluating participants' feelings of safety when interacting with an AV, and assessing the efficiency/receptivity/learning effect of different eHMI designs. They also provide knowledge on choosing the most appropriate time gaps and conditions for testing future eHMIs.

For example, in a study by [16], participants saw a pair of vehicles approaching from the right, and were asked to cross the road, after the first vehicle had passed. The approaching speed of the second vehicle was manipulated (25 mph, 30 mph or 35 mph) and the time gap between the two vehicles ranged between 1–8 s (with 1 s increments). In addition, the second vehicle was either decelerating as it approached the pedestrian, or not. Data from the decelerating trials showed that 51% of crossings happened before the second vehicle decelerated, and 31% of crossings happened after the approaching vehicle had stopped, with only 18% of crossings happening during the deceleration. Results which are also confirmed by the modelling work of [31] in this context.

Previous Wizard-of-Oz studies investigating pedestrian response to "fake AVs", have shown that pedestrians did not feel comfortable, or safe, crossing the road in front of the specially customised vehicle, where the driver (sitting behind a fake steering wheel in the passenger seat of the modified vehicle) was seen to be asleep or deeply engaged in reading a newspaper [17]. In the interACT project, we investigated this matter further, using an HMD VR-based study, where participants were asked to cross the road in a set-up similar to [16] described above (see [19]). To establish if driver presence and attention affected crossing behaviour, the second vehicle in this study (which was always travelling at 30 mph) was presented in three different conditions (no driver, distracted driver – looking down, and attentive driver – looking straight ahead). To ensure the VR setup was realistic, and the drivers were actually visible, all participants completed a short set of trials at the end of the experiment, pressing a button on the controller to confirm the presence or absence of drivers in the vehicle. Although pedestrian crossing behaviour was not affected by the three conditions, follow-on questionnaires on perceived behavioural control and perceived risk [18], showed that the "driver present" conditions were rated higher than driver distracted/driver absent trials [19].

Finally, the interACT project's VR studies on explicit communication by AVs, which have utilised various visual message strategies such as ground projections, directed signal lamps and LED bands, have thus far shown a reduced initiation time for pedestrians to cross the road. Although objective studies showed little difference in crossing initiation time between the different concepts, participants reportedly preferred animations and symbols to static images [20] (Fig. 2).

Fig. 2. Depiction of one of the VR studies, showing participant with the HMD and the road environment used for studying crossing behaviour

4 Computational Models

4.1 Neurobiologically-Informed Mathematical Models

Another, even more concrete, way of describing how road users behave when they interact with each other in shared space, is to develop mathematical models that permit computer simulations of the interaction behaviour. Traffic microsimulation is a well-established field of research, and commercial software products exist that permit traffic simulations that are accurate on the scale of a large junction or a city centre, for example, to predict how a range of alternative road infrastructure designs will affect traffic throughput [21, 22]. The road user behaviour models in these traffic simulations are, however, not designed to capture the details of local interactions, and this underdeveloped area is now garnering increasing attention, with some modellers approaching it from a traffic microsimulation starting point [24, 25], and others addressing it as a data-driven machine learning challenge [26–28].

In the interACT project, a third type of approach has been taken, partly because the aim has been very specific: to generate useful insights and tools for the AV-human interaction design work in the project. To this end, a novel modelling framework has been proposed [29], building on psychological and neuroscientific models of decision-making, such as evidence accumulation [23, 29, 30]. The benefit of this type of framework is that it allows the model to integrate sensory evidence, both from AV movement and eHMI messages, in a manner that is both mathematically straightforward, and neuro-biologically plausible. This model has been applied to a pedestrian crossing decision, qualitatively reproducing the empirically observed tendency of pedestrians to either cross early in front of a yielding vehicle, or otherwise wait until the vehicle has come to a complete stop [16, 31]. This type model can also be used to study efficiency of AV-human interactions [31]. Figure 3 shows how this tentative model predicts a considerable traffic flow benefit, both of the AV providing an eHMI message which signals yielding (panel b compared to panel a), and of the AV slightly exaggerating its yielding deceleration (moving upward along the y axis). Currently ongoing, but not yet published work, has shown that this type of model can be successfully fitted to observed human behaviour, both in vehicle-pedestrian and vehicle-vehicle scenarios.

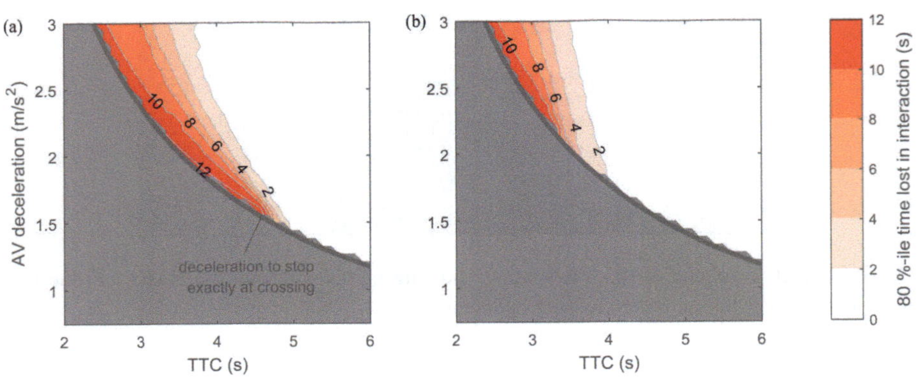

Fig. 3. Results of simulations with the pedestrian crossing model proposed by [31], showing how the 80 percentile of lost time for the AV due to the interaction (i.e., how much earlier the AV would have arrived at its destination, had the pedestrian not been present), as a function of time left to the pedestrian crossing when the vehicle initiates yielding (TTC), and the magnitude of the yielding deceleration. Panels (a) and (b) show results without and with an eHMI indication of yielding, respectively. Figure from [31]. Copyright © 2018 National Academy of Sciences. Reprinted by Permission of SAGE Publications, Inc.

4.2 Using Game Theory to Understand the Interaction Between Pedestrians and AVs

Controlling autonomous vehicles in the presence of pedestrians, when they are competing for the same space, requires an understanding of the processes of interaction and negotiation between them. Game theory provides a formal basis for modelling multi-agent competitive interactions. For instance [32] constructed a mathematical model of interactions between two such agents approaching an unmarked intersection, a technique which can be modified to a range of scenarios, such as pedestrians crossing the road, vehicle-vehicle interactions, and pedestrian-pedestrian interactions. The model is based on the "game of chicken" of Game Theory, extended to a temporal model, and is deliberately simplified as much as possible, to illustrate the core idea, using heavy quantification of both space and time. Here, at each "tick", a game is played in discrete time, in which the two agents can choose to yield, by moving forward slowly, or asserting themselves by moving forward quickly. They approach each other in this way, and will collide, unless one of them yields. The mathematics of the model shows that the optimal strategy when both players have the same utilities, is the same for both players, and is probabilistic. At each time, they should flip a biased coin and yield if they get heads; with the bias of the coins increasing towards heads with certainty, as time runs out. The incentive to cooperate and avoid a disaster thus grows with time. The model further shows that if the utilities are modified to make one player survive better in a collision – such as being a vehicle vs a pedestrian, or an SUV vs a smaller car – then even a small change in these utilities will break the symmetry of the model, and give them a relatively high chance of winning every interaction.

The above model has been tested in experimental laboratory settings e.g. [33], where a board game was played by seated participants, to model the same collision scenario. This study showed that the behaviour of the players can be fitted via a Gaussian Process over parameters, using the model. The authors then extended this setup to a more realistic, but still heavily constrained, physical laboratory experiment, with participants walking towards each other in discrete time and space, using the same methods to fit the model parameters.

Overall, results from this model suggests that AVs must retain an ability to deliberately cause harm of some sort to other road users, in order to make any progress at all. The value of this model has been included in a consultation paper, currently out in circulation by the UK Law Commission, on Autonomous Vehicles [34], and may contribute to changing the law of the UK, to ensure a fairer relationship between AVs and other road users.

It can be argued that the "chicken model" described above is quite crude in its assumptions, especially as it does not yet include any ability for the players to signal information to each other than via their speeds and positions. However, it has been argued that real-world interactions may include many such signals, such as eye contact, head direction, and body language, which, as shown in our own observations is particularly useful for solving conflicts. To begin to form an understanding of these signals for later use in the chicken model, we used the video recordings from Leeds outlined in Sect. 2.2, to investigate which of a bank of such signals are useful to predict the final outcome of pedestrian-car interactions [35].

This basic model was later extended to a temporal filtration model, which shows how the probability of the game winner evolves over time as a function of the signals provided [36]. Future work will try to integrate these models, using signalling behaviour as an additional input to the game theory mathematics, to refine its solutions, then test them in virtual reality, and physical AV human experiments.

5 Summary and Conclusions

This paper provides an overview of the complex relationship that exists between different road users in a mixed traffic environment, and summarises the preliminary results of a set of behavioural studies conducted in this context, as part of the European interACT project. It highlights the value of using different methodologies to understand the behaviour of road users in current traffic settings, illustrating the complexity of road user behaviour, and the influence of infrastructural, social, and cultural norms, in this context. Knowledge is currently building on how new methods can be used to help towards innovation of new forms of communication and interaction for future AVs, such as the value of external Human Machine Interfaces to replace the communication currently provided by human drivers. The challenge here is to ensure that AVs' manoeuvres in mixed settings are safe, and therefore acceptable by all road users. However, it is also important to ensure that AVs' behaviour and progress is not restricted by their, currently limited, obstacle detection rules, which will reduce their ability to achieve a smooth and uninterrupted journey. As more AVs are introduced for testing in such settings, it is likely that human interactions with them will also change, following some level of Behavioural Adaptation [37]. This allows the likelihood of more knowledge to be gained by both AV developers, and human road users, ensuring that these two actors can cooperate more efficiently with each other in the future urban environment.

References

1. Merat, N., Louw, T., Madigan, R., Wilbrink, M., Schieben, A.: What externally presented information do VRUs require when interacting with fully automated road transport systems in shared space? Accid. Anal. Prev. **118**, 244–252 (2018). https://doi.org/10.1016/j.aap.2018.03.018
2. Lee, Y.M., Madigan, R., Giles, O., Garach-Morcillo, L., Markkula, G., Fox, C., Camara, F., Rothmueller, M., Vendelbo-Larsen, S.A., Rasmussen, P.H., Dietrich, A., Nathanael, D., Portouli, V., Schieben, A., Merat, N.: What can we learn from current communication and interactions by European road users: a contribution to external interface design for future Automated Vehicles (submitted)
3. International Standards Organisation committee TC22/WC 39
4. Dietrich, A., Ruenz, J.: Observing traffic–utilizing a ground based LiDAR and observation protocols at a T-junction in Germany. In: Bagnara, S., Tartaglia, R., Albolino, S., Alexander, T., Fujita, Y. (eds.) Proceedings of the 20th Congress of the International Ergonomics Association, IEA 2018. Advances in Intelligent Systems and Computing, 823. Springer, Cham (2019). https://doi.org/10.1007/978-3-319-96074-6_56

5. Dietrich, A., Bengler, K., Portouli, E., Dimitris, N., Johannes, R., Jingyuan, W., Merat, N., Madigan, R., Lee, Y.M., Markkula, G., Giles, O., Fox, C., Camara, F.: interACT D.2.1. Preliminary description of psychological models on human-human interaction in traffic (2018) https://www.interact-roadautomation.eu/wp-content/uploads/interACT_WP2_D2.1-_PsychologicalModels_v1.0_uploadwebsite-1.pdf
6. Dietrich, A., Willrodt, J.-H., Wagner, K., Bengler, K.: Projection-based external human machine interfaces–enabling interaction between automated vehicles and pedestrians. In: Proceedings of the DSC 2018 Europe VR, Driving Simulation & Virtual Reality Conference & Exhibition, pp. 43–50 (2018)
7. Risto, M., Emmenegger, C., Vickhuyzen, E., Cefkin, M., Hollan, J.: Human-vehicle interfaces: the power of vehicle movement gestures in human road user coordination. In: Proceedings of the 9th International Driving Symposium on Human Factors in Driver Assessment, Training and Vehicle Design, pp. 186–192 (2017)
8. Schieben, A., Wilbrink, M., Kettwich, C., Madigan, R., Louw, T., Merat, N.: Designing the interaction of automated vehicles with other traffic participants: design considerations based on human needs and expectations. Cogn. Technol. Work **109**, 1–17 (2018)
9. Sucha, M., Dostal, D., Risser, R.: Pedestrian-driver communication and decision strategies at marked crossings. Accid. Anal. Prev. **102**, 41–50 (2017)
10. Dey, D., Terken, J.: Pedestrian interaction with vehicles: roles of explicit and implicit communication. In: Proceedings of the 9th International Conference on Automotive User Interfaces and Interactive Vehicular Applications (Au-tomotiveUI'17), pp. 109–113 (2017). http://doi.org/10.1145/3122986.3123009
11. Lagstorm, T., Lundgren, V.M.: AVIP-autonomous vehicles interaction with pedestrians. Master's thesis, Chalmers University of Technology, Gothenborg, Sweden (2015)
12. Clamann, M., Aubert, M., Cummings, M.L.: Evaluation of vehicle-to-pedestrian communication displays for autonomous vehicles. Technical report (2017)
13. Habibovic, A., Andersson, J., Nilsson, M., Lundgren, V.M., Nilsson, J.: Evaluating interactions with non-existing automated vehicles: three Wizard of Oz approaches. In: IEEE Intelligent Vehicles Symposium (IV), Gothenburg, pp. 32–37 (2016). https://doi.org/10.1109/IVS.2016.7535360
14. Deb, S., Strawderman, L.J., Carruth, D.W.: Investigating pedestrian suggestions for external features on fully autonomous vehicles: a virtual reality experiment. Transp. Res. Part F **59**, 135–149 (2018)
15. Sorokin, L., Chadowitz, R., Kauffmann, N.A.: Change of perspective: designing the automated vehicle as a new social actor in a public space. In: CHI'19 Extended Abstracts, Glasgow, Scotland, UK, 4–9 May 2019. https://doi.org/10.1145/3290607.3299044
16. Lee, Y.M., Uttley, J., Solernou, A., Giles, O., Markkula, G., Romano, R., Merat, N.: Investigating pedestrians' crossing behaviour during car deceleration using wireless head mounted display: an application towards the evaluation of eHMI of automated vehicles, New Mexico, Driving Assessment (2019)
17. Lundgren, V.M. Habibovic, A., Andersson, J., Lagström, T., Nilsson, M., Sirkka, A., Fagerlönn, J., Fredriksson, R., Edgren, C., Krupenia, S., Saluäär, D.: Will there be new communication needs when introducing automated vehicles to the urban context? In: Stanton, N., Landry, S., Di Bucchianico, G., Vallicelli, A. (eds.) Advances in Human Aspects of Transportation, Advances in Intelligent Systems and Computing, 484, Springer, Cham (2017). https://doi.org/10.1007/978-3-319-41682-3_41
18. Zhou, R., Horrey, W.J., Yu, R.: The effect of conformity tendency on pedestrians' road-crossing intentions in China: an application of the theory of planned behaviour. Accid. Anal. Prev. **41**(3), 491–497 (2009)

19. Velasco, J.P., Lee, Y.M., Uttley, J., Solernou, A., Farah, H., van Arem, B., Hagenzieker, M., Merat, N.: Interactions with automated vehicles: the effect of drivers' attentiveness and presence on pedestrians' road crossing behaviour. In: Road Safety and Simulation Conferences, University of Iowa (2019, submitted)
20. Dietrich, A., Tondera, M., Bengler, K.: Automated vehicles in urban traffic: the effects of kinematics and eHMI on pedestrian crossing behaviour. In: Road Safety and Simulation Conferences, Universty of Iowa (2019, submitted)
21. Helbing, D.: Traffic and related self-driven many-particle systems. Rev. Mod. Phys. **73**(4), 1067–1141 (2011). https://doi.org/10.1103/RevModPhys.73.1067
22. Casas, J., Perarnau, J., Torday, A.: The need to combine different traffic modelling levels for effectively tackling large-scale projects adding a hybrid meso/micro approach. Procedia Soc. Behav. Sci. **20**, 251–262 (2011)
23. Ratcliff, R., Smith, P.L., Brown, S.D., McKoon, G.: Diffusion decision model: current issues and history. Trends Cogn. Sci. **20**(4), 260–281 (2016)
24. Rinke, N., Schiermeyer, C., Pascucci, F., Berkhahn, V., Friedrich, B.: A multi-layer social force approach to model interactions in shared spaces using collision prediction. Transp. Res. Procedia **25**, 1249–1267 (2017)
25. Anvari, B., Bell, M.G., Sivakumar, A., Ochieng, W.Y.: Modelling shared space users via rule-based social force model. Transp. Res. Part C: Emerg. Technol. **51**, 83–103 (2015)
26. Behbahani, F., Shiarlis, K., Chen, X., Kurin, V., Kasewa, S., Stirbu, C., Gomes, J., Paul, S., Oliehoek, F.A., Messias, J., Whiteson, S.: Learning from demonstration in the wild (2018). arXiv preprint arXiv:1811.03516
27. Cheng, H., Sester, M.: Mixed traffic trajectory prediction using LSTM–based models in shared space. In: The Annual International Conference on Geographic Information Science, pp. 309–325 (2018)
28. Chen, B., Zhao, D., Peng, H.: Evaluation of automated vehicles encountering pedestrians at signalized crossings. In: 2017 IEEE Intelligent Vehicles Symposium (IV), pp. 1679–1685, June 2017
29. Markkula, G., Boer, E., Romano, R., Merat, N.: Sustained sensorimotor control as intermittent decisions about prediction errors: computational framework and application to ground vehicle steering. Biol. Cybern. **112**(3), 181–207 (2018)
30. Purcell, B.A., Heitz, R.P., Cohen, J.Y., Schall, J.D., Logan, G.D., Palmeri, T.J.: Neurally constrained modeling of perceptual decision making. Psychol. Rev. **117**(4), 1113–1143 (2010)
31. Markkula, G., Romano, R., Madigan, R., Fox, C.W., Giles, O.T., Merat, N.: Models of human decision-making as tools for estimating and optimizing impacts of vehicle automation. Transp. Res. Rec. (2018). https://doi.org/10.1177/0361198118792131
32. Fox, C., Camara, F., Markkula, G., Romano, R.A., Madigan, R., Merat, N.: When should the chicken cross the road? Game theory for autonomous vehicle-human interactions. In: Proceedings of the 4th International Conference on Vehicle Technology and Intelligent Transport Systems (VEHITS), Funchal, Madeira, 16–18 March 2018
33. Camara, F., Romano, R., Markkula, G., Madigan, R., Merat, N., Fox, C.: Empirical game theory of pedestrian interaction for autonomous vehicles. In: Proceedings of the Measuring Behaviour 2018: International Conference on Methods and Techniques in Behavioural Research (2018)
34. Automated Vehicles: Summary of the preliminary consultation paper. https://www.lawcom.gov.uk/project/automated-vehicles/consultation

35. Camara, F., Giles, O., Madigan, R., Rothmueller, M., Rasmussen, P.H., Vendel-bo-Larsen, A., Markkula, G., Lee, Y.M., Merat, N., Fox, C.W.: Predicting pedestrian road-crossing assertiveness for autonomous vehicle control. In: Proceedings of the 21st IEEE International Conference on Intelligent Transportation Systems (2018)
36. Camara, F., Giles, O., Madigan, R., Rothmuller, M., Rasmussen, P.H., Vendel-bo-Larsen, S. A., Markkula, G., Lee, Y.M., Garach-Morcillo, L., Merat, N., Fox, C.: Filtration analysis of pedestrian-vehicle interactions for autonomous vehicle control. In: 15th International Conference on Intelligent Autonomous Systems, Baden-Baden, Germany, 11–15 June 2018
37. Sullivan, J.M., Flannagan, M.J., Pradhan, A.K., Bao, S.: Literature review of behavioural adaptation to advanced driver assistance systems. AAA Foundation for Traffic Safety, Washington, DC (2016)

The Role of Human Factors in the Design of Automated Vehicle External Communication

W. Andy Schaudt(✉), Sheldon Russell, and Justin M. Owens

Virginia Tech Transportation Institute, 3500 Transportation Research Plaza, Blacksburg, VA, USA
{aschaudt, srussell, jowens}@vtti.vt.edu

Abstract. This chapter presents a summary of the 2018 AVS Breakout Session 25, The Role of Human Factors in the Design of Automated Vehicle External Communications. The session was scheduled for four hours with the majority of the time dedicated to presentations from three speakers and the remaining for an interactive exercise. The three speakers presented on a variety of topics, which included automated vehicle research projects involving external communication, studies investigating vulnerable road user behavior, as well as activities underway exploring the potential value of international standardization. Key points included the importance of exploring new metrics for measuring the performance of automated vehicle external communication, the need to study these systems longitudinally, and the importance of investigating the relationship between safety and road user trust and acceptance of these systems.

Keywords: Automated vehicles · Highly automated vehicles · External communication · Human factors · External signals · Human-machine interface · Vulnerable road user

1 Introduction

Research is underway across numerous countries investigating the potential of new external communication methods for highly automated vehicles (HAVs). As drivers are replaced, we lose human-to-human communication between road users such as eye contact and hand gestures. With this loss, and our uncertainty as to whether HAVs will be able to perceive and communicate intent in the same ways that a human can, do we need a new form of HAV external communication? If so, how and what should it communicate to other road users?

The 2018 AVS Breakout Session that this chapter is based on continued upon the preceding 2017 AVS Breakout Session by providing an update on relevant activities in research and standardization [1]. To accomplish this, the organizers invited experts to provide brief presentations on their research. In addition, there was a breakout exercise performed to explore novel solutions, unique use cases, and to discuss the human factors design implications. In Sect. 2 we present summaries of the presentations by invited

G. Meyer and S. Beiker (Eds.): AUVSI 2019, LNMOB, pp. 34–38, 2019.
https://doi.org/10.1007/978-3-030-22933-7_4

experts. Section 3 will describe the exercise that was performed by all those in attendance. Finally, Sect. 4 will provide highlights of the discussion that was generated.

2 Current Research in HAV External Communication

2.1 Presentation Summaries

2.1.1 Study of Automated Vehicle External Communication in the Wild

Andy Schaudt from the Virginia Tech Transportation Institute (VTTI) presented on a completed study in partnership with the Ford Motor Company. This study was conducted on the public roadways of Arlington, VA. The purpose of this study was to evaluate road user behavior in the presence of candidate automated vehicle external communication visual signals. Road users of interest included other drivers and vulnerable road users (e.g. pedestrians, motorcyclists). This field study deployed a seemingly driverless vehicle (driver hidden in seat suit) without visual external communication (baseline) and with visual external communication (intervention). Over 2,350 miles (167 h) of video and kinematic data were collected over a 30-day period. The primary dependent variable of interest was whether a change in other road user behavior was observed for those who noticed the vehicle between baseline and intervention conditions. Data were collected across a variety of scenarios such as driving straight or turning through intersections (both signalized and signage only), roadway speeds up to 45 mph, mock passenger drop-offs and pickups, etc. In addition to video and kinematic data collected, intercept surveys with pedestrians were also delivered. The purpose of these were to understand people's understanding of the signal types, and also gauge their level of trust and acceptance.

Preliminary results were presented, indicating that there were no statistically significant differences of behavior change between baseline and intervention conditions. However, surveys did indicate some positive results that pedestrians believe these types of signals would provide value.

2.1.2 Road User Behavior in the Presence of Automated Vehicle External Communication in Northern California

Jingyi Zhang from the Nissan Research Center presented on results from a simulator study conducted in Northern California of communication intent of multiple autonomous vehicles. The focus of the study was to understand the effects of external communication on drivers' interactions, understanding, and feedback, as well as distraction potential. Results indicated that the intention indicators would not cause additional distractions. In addition, people liked the concept of communicating vehicle intent and/or motion (although different types of road users need to be investigated separately). Finally, preferences on color and type of signal (e.g. icon, light) can vary. Preliminary findings indicate that light configurations were equally recognizable as icon configurations, but possibly harder to interpret on the first exposure.

2.1.3 Updates on ISO Activity Surrounding External Communication

Andy Schaudt from VTTI presented on current activities from the International Organization for Standardization (ISO) regarding HAV external communication. Mr. Schaudt first described a technical report currently in the publication process titled "Ergonomic aspects of external visual communication from automated vehicles to other road users". The purpose of this document is to provide guidance for developers of automated vehicle visual external communication systems. Road users will need to understand how to safely interact with these vehicles, and if external communication will be deployed, it is important to start working on a common approach. Consistency across the automotive industry will minimize potential road user confusion and may have the added benefit of establishing societal trust with respect to HAVs.

Mr. Schaudt also described two new work item proposals in progress from the Road Vehicle Ergonomics subcommittee. The first proposal will be an AV external communication general design guidance. The second proposal will be an AV external communication experimental evaluation guidance with coding schemes.

2.1.4 Human Factors Considerations for Enabling Safe Interactions Between AVs & VRUs

Dr. Justin Owens from VTTI presented on opportunities for enabling safe interactions between HAVs and vulnerable road users (VRUs). He identified potential safety benefits of HAVs if designed with human factors considerations. These included the potential for improved perception and reaction times, the elimination of risk associated with distraction and impairment, and the opportunities of designing better affordances for people with disabilities. Although there are many opportunities, Dr. Owens also addressed some human factors challenges. Such as, how do vehicles and pedestrians communicate control and intent? Will, and should, people be able to properly identify the different levels of driving automation? How will future infrastructure support these interactions? Finally, Dr. Owens also described ongoing VRU research involving VTTI. One such example is the development and evaluation of a VRU mobility assistance platform funded by the Center for Advanced Transportation Mobility, a Tier 1 UTC, that will enable personalized non-driving navigation directions of the built environment for people who have movement disabilities.

3 Breakout Exercise

Organizers prepared a breakout exercise based around unique use cases and types of HAV external communication that would help generate interaction and discussion among attendees on the human factors implications. Attendees were instructed to breakout into five separate groups, select a use case, and then identify and discuss the human factors considerations. Prior to the introduction of use cases, the organizers first provided a list of external communication types previously identified by the ISO Road Vehicle Ergonomics subcommittee that could be used during the exercise:

- Vehicle State (e.g. tones and beeps as truck backs up)
- Driving Mode (e.g. automation level)

- Perception (e.g. HAV tells vehicle following behind about pedestrians ahead)
- Recognition & Acknowledgement (e.g. communicates that it recognizes a pedestrian)
- Belief State (e.g. communicates braking distance to encroaching pedestrian)
- Guidance (e.g. tells pedestrian when safe to cross)
- Intent (e.g. intends to yield).

3.1 Use Case A: Self-driving Taxi Performing a Passenger Pickup

The purpose of Use Case A was to provide a common example of a taxi approaching a passenger to pick up and transport. Groups were provided a series of probing questions to help generate discussion about human factors implications on external communication design. For Use Case A, these questions were:

- Who are the relevant road users involved?
- What types of external communication might be relevant for this use case?
- How should the HAV communicate in order for the signal to be clear and learnable?
- Will the proposed signal add or remove confusion?
- Will this type of communication fit within the current roadway infrastructure design?

3.2 Use Case B: HAV Transporting Goods

The purpose of Use Case B was to provide an example of a common roadway scenario of a vehicle transporting goods (e.g. package delivery) except with no driver or rider. The same questions used in Use Case A were used for Use Case B.

3.3 Use Case C: Remotely Operated HAV

The purpose of Use Case C was to provide an example of an uncommon scenario where a vehicle is being operated remotely transporting one or more passengers. The same questions used in Use Case A and Use Case B were used for Use Case C.

4 Discussion and Research Needs

In each breakout group, attendees engaged in substantial discussion. This discussion resulted in some outstanding questions and research needs for future consideration. Highlights from the discussion are presented below:

- Although self-driving taxi users will likely hail and communicate via mobile app, how will customers communicate with self-driving delivery vehicles? For example, do they meet the vehicle at the street? If so, when? How is this more convenient than a driver bringing the package to your front door?
- Should self-driving delivery vehicles communicate that they make stops more frequently?

- Should a self-driving delivery vehicle communicate guidance to following-vehicle drivers when it is safe to go around them (e.g. pass)?
- Will self-driving vehicles double-park like many current delivery vehicles do in urban environments? Or will dedicated areas for drop-offs and pickups be established? What signals will be used during this event?
- Should a remotely operated HAV identify itself differently than standard HAV?
- Should a remotely operated taxi communicate its operation status to riders? What if the vehicle switches back and forth between self-driving and remotely operated?
- Should self-driving Taxis communicate their high occupancy vehicle (HOV) status for utilization of HOV lanes?
- How can a cell phone be leveraged for communicating information to passengers with disabilities (e.g. onboarding)? Should there be a common design adopted by car manufacturers?

This breakout session not only included presentations from experts currently researching the topic of HAV external communication, but also provided an interactive exercise using specific use cases to generate substantial discussion towards identifying new research questions not yet explored. Some important takeaways included the need to study these systems over longer periods of time, the importance of investigating the relationship between safety and road user trust and acceptance of these systems, and finally that special consideration should be given to the design of external communication based on the utility of the HAV.

Reference

1. Schaudt, W.A., Russell, S.: Judging a car by its cover: human factors implications for automated vehicle external communication. In: Meyer, G., Beiker, S. (eds.) Road Vehicle Automation 5. Springer International Publishing AG (2018). ISBN 978-3-319-94896-6

Designing Automated Vehicles Around Human Values

J. Christian Gerdes[1], Sarah M. Thornton[1], and Jason Millar[2](✉)

[1] Stanford University, 416 Escondido Mall, Building 550, Room 136, Stanford, CA 94305, USA
gerdes@stanford.edu, smthorn@alumni.stanford.edu
[2] University of Ottawa, 5026C-800 King Edward Ave, Ottawa, ON K1N 6N5, Canada
jmillar@uottawa.ca

Abstract. The impact of automated vehicles will reverberate across society in many dimensions, changing our expectations of mobility, safety, employment and other aspects of life we value. These major societal changes will, in turn, be the result of a number of small engineering decisions that, when aggregated, determine the system behavior. For automated vehicles to have the benefits their advocates envision, we must bridge the gap between these individual decisions and the societal impacts they create. This paper discusses some of the challenges faced by engineers in bridging this gap and proposes a value-centered approach to the design of automated vehicles. Such an approach engages stakeholders early in the process, identifying values and tensions with enough specificity to drive subsequent engineering choices.

Keywords: Autonomous vehicles · Human values · Ethical programming

1 Introduction

Have you ever driven down a street you have never driven down before because you were guided there by a navigation app? With in-car navigation systems and cell phones, such experiences have become a regular aspect of driving for most Americans. In our quest for reduced travel time and less stressful driving, we have delegated the choice of what streets we travel to routing algorithms that can account for factors such as real-time traffic about which we can only guess.

Yet, while drivers appreciate the time and stress saved, neighborhoods now have to contend with increased traffic. Streets that were once only known to local residents are now regular commute routes, changing the character of the neighborhoods through which they wind. Stories abound of quiet streets turned into "speedways" by the presence of these new road users [1, 2].

This battle over neighborhood streets is a classic example of a value tension. There is an obvious conflict between travelers wanting the shortest driving time and neighborhoods wanting to preserve their relative tranquility. It is easy to argue that increased traffic through residential neighborhoods with children, pedestrians and bicycles poses

G. Meyer and S. Beiker (Eds.): AUVSI 2019, LNMOB, pp. 39–48, 2019.
https://doi.org/10.1007/978-3-030-22933-7_5

an increased threat to safety. But, similarly, roads are a public good. Shouldn't they be used as efficiently as possible? When faced with such tensions, who should decide?

In the past, these competing demands were balanced by traffic engineers using standards for road design that were themselves the result of decades of research and public comment. Increasingly, however, these decisions are made by the engineer who designs the routing algorithm, often without much awareness of the values implicated by that choice or the tensions that might arise as a result. What is most striking is how quickly we have moved from road usage determined by standards to road usage determined by weights in a routing algorithm. As the reach of systems like navigation apps expands and automated vehicles take to the streets, small engineering decisions will increasingly, and rapidly, produce large societal impacts. Instead of societal aspects such as road safety or traffic being the aggregate of many human decisions or carefully crafted standards, huge swaths of our society will hinge on small engineering decisions.

In such a future, it becomes critical to center engineering decisions around the human values and societal characteristics impacted by those decisions. But what exactly are human values? Borning and Muller [3] define "value" as "what a person or group of people consider important in life." When considering transportation, mobility is an important human value but far from the only one. The human act of driving requires a constant balancing of competing values such as mobility, safety and legality, the results of that balancing act differing according to how each driver handles the many driving scenarios encountered in a single trip. This balancing act can be complicated. Depending on the scenario, one value may clearly take precedence over another, such as when drivers decide to respect a stop sign at a busy intersection—here, safety trumps mobility.

In designing automated vehicles, engineers cannot avoid making decisions that implicate safety, mobility and legality. Several designers of motion planning algorithms have accordingly sought to address and deliberately resolve value tensions with their algorithm designs. Bouton, Nakhaei, Fujimura, and Kochenderfer [4] explicitly balanced mobility and safety in the design of a motion planning algorithm for navigating occluded scenarios by penalizing collisions and rewarding the vehicle for completing maneuvers. Wongpiromsarn, Karaman, and Frazzoli [5] encoded traffic laws in order to synthesize a motion planning algorithm that navigates a two-lane roadway with an obstacle and a double yellow line. The autonomous vehicle is able to navigate around the obstacle because the traffic rules are encoded such that crossing a double yellow is allowable after the vehicle first comes to a stop behind the obstacle, thus balancing legality, safety and mobility.

These examples show how engineers can incorporate or resolve value tensions when those tensions are clear from the start. The fact that humans care about values, and that engineers can incorporate identified values in their designs, motivates us to consider designing autonomous vehicles with human values in mind. The next crucial piece is to consider *how* to go about designing with human values in mind. In the following sections, we present a value-centered approach to designing automated vehicles that focuses the design process on human values from the beginning. In what we propose, human values serve as common ground in all discussions of automated vehicle design across stakeholder groups. While resolving tensions between different values can be challenging, starting the discussion from the standpoint of values and having meaningful conversations about those tensions need not be. Human values are vitally important; where we strive for our designs to function well in society, it makes sense to situate values at the core of engineering decision-making.

2 Some Value Tensions with Automated Vehicles

To further illustrate the need for considering human values and the challenges engineers may face when attempting to do so, we detail a few examples of design problems that can arise during the development of an automated vehicle. These examples not only cover different aspects of automation but also three different challenges to a value-centered design approach: identifying the underlying values, resolving tensions among those values and acknowledging scenario-dependent immediacy effects.

2.1 Routing Algorithms

A foundational feature of an automated vehicle is to deliver occupants (or goods) from point A to point B. In order to accomplish this feature, a route must be planned for the automated vehicle to follow. A popular routing algorithm is a shortest path search algorithm known as A* (pronounced A-star) [6], which considers both cost and heuristics for traversing a road segment. There are many factors that can be included in the creation of the cost and heuristics. For example, speed limit and length of the road segment could be included in the cost calculus. The heuristic may be a rough guide, like the Manhattan distance, in order to encourage exploration of the search space in the direction of the target destination.

From the standpoint of an AV programmer, it makes perfect sense to consider routing in terms of the travel time for the occupants. But, as discussed in the Introduction, this is only one of many values associated with vehicle routing. Vehicle routing affects the tranquility of neighborhoods, the environmental impact of the transportation system and, in a mobility-on-demand system, waiting times and service quality. Safety may vary across potential routes, with routes through areas with high concentrations of vulnerable road users decreasing safety for those outside the vehicle and routes through areas of high crime potentially decreasing safety for vehicle occupants. Should these dimensions be added to routing algorithms for AVs or does doing so raise the prospects of neighborhoods being redlined and unable to share in the benefits AVs will provide? Routing algorithms demonstrate how difficult it can be for engineers to properly scope the range of human values relevant to a design choice in isolation. Stakeholder engagement to form an inclusive and diverse understanding of those values becomes critical.

2.2 Vehicle Platooning

One approach to automating vehicles involves creating platoons of vehicles that can follow each other on the highway at close spacing, as depicted in Fig. 1. In the case of heavy trucks, the focus of such systems has been to reduce fuel consumption [7] and, through coordinated braking, reduce the likelihood of collision in certain accident scenarios. There are many values to consider beyond these improvements in fuel consumption and safety, however. Other road users need to be able to maneuver around or between members of a platoon, particularly at highway exits. When both trucks are operated by human drivers, visibility becomes a key aspect of both safety and human comfort. Furthermore, there is a central tension between the two main objectives of the system. Traveling at a closer spacing can improve fuel economy but raise safety concerns about the vehicles colliding with each other due to, for instance, different braking capability.

Fig. 1. Two heavy truck vehicles platooning on the highway (photo: Peloton Technologies)

The challenge for engineers designing platooning systems is not to surface these values or tensions; all of these are clear from a single platooning experience on the highway. Rather, the challenge is that all of these values hinge on a single decision – the choice of following distance. While this number is trivial to set or modify in a following algorithm, all aspects of system performance and public acceptance stem from that single choice. Platooning is therefore a great example of the need to develop ways of prioritizing or balancing values when so many values are determined by a single design parameter.

2.3 Pedestrian Interactions

Automated vehicles navigating urban environments must determine the appropriate speed at which to approach pedestrian crosswalks. In the California Vehicle Code, §21950 states that drivers must yield the right of way to any pedestrian within a marked crosswalk. When the pedestrian is actually in the crosswalk, the law's requirement and the necessary vehicle action are rather straightforward. But in order to respect this law and ensure pedestrians' safety, automated vehicles must also anticipate whether or not the pedestrian will be in the crosswalk at the time the vehicle arrives. As human drivers know, such predictions can be far from trivial. Figure 2 depicts a pedestrian standing in an influential area next to the crosswalk. The pedestrian is completely stopped while looking down at a device he is holding, presenting an ambiguous situation for the automated vehicle. While sometimes this ambiguity resolves as the vehicle gets closer, in this case it does not and the AV must choose a speed through the intersection accordingly. The value tension manifests as conflicts over mobility of different road users and among safety, mobility and legality in setting a speed.

Fig. 2. An automated vehicle approaches a pedestrian crosswalk. At approximately 10 m away (left), it is difficult to discern the intent of the pedestrian. Closer to the crosswalk (right), it is still difficult to discern the intent of the pedestrian.

Viewing this as a tension between the vehicle and the pedestrian alone, however, obscures other value tensions that arise. While legal concerns and the immediacy of the pedestrian may prompt the engineer to frame this design problem narrowly, and take a very conservative approach to speed setting in these cases, such choices impact a broader set of indirect road users as well. Vehicles behind the AV may see their mobility unnecessarily decreased if the algorithm is too conservative, while the broader effects could be an unacceptable decrease in overall systemwide mobility. Furthermore, some human drivers may attempt to overtake the AV as a result, increasing other risky behaviors overall. Crosswalks present an example where narrowly considering value tensions of the immediate stakeholders may fail to adequately reflect the values of others impacted by these decisions.

These three examples are just a sampling of the value tensions that arise when designing automated vehicles, a technology that will alter society in myriad dimensions across an array of stakeholder groups. While these examples highlight different challenges in value-centered design, they share a common theme: all of these tensions involve what seem like rather mundane decisions. In fact, the societal impact of automated vehicles will ultimately be determined by a vast number of seemingly small engineering decisions. The ethical design of automated vehicles, therefore, requires that these decisions be grounded in human values.

3 Ethical Programming, Not Programming Ethics

This focus on the broader impact of small engineering decisions runs counter to the way that ethics is often discussed with respect to automated vehicles. In the past few years, the topic of ethics for automated vehicles has often been equated to finding solutions for the infamous "trolley problem" arising from philosophy. The trolley problem, first invented as a philosophical thought experiment, asks one to consider a scenario in which an uncontrollable trolley is barreling unstoppably down a set of railway tracks because of its broken brakes. If the trolley continues on its way, it is guaranteed to kill five unsuspecting people on the tracks ahead. You (a bystander) are standing by a switch, giving you the power to intervene and divert the trolley to another set of tracks on which there is a single unsuspecting person that would surely perish [8]. Moral psychologist Greene [9] has demonstrated that variations of the trolley problem can provide deep insights into the way the human brain processes these ethical dilemmas.

It's easy to replace the trolley with an autonomous vehicle and imagine that an algorithm would need to determine whether to continue on the collision course with five pedestrians or swerve and kill one other. Given that this would be a tragic scenario for an autonomous vehicle (or anyone) to encounter on the roadway, the trolley problem, imagined as a real-world hypothetical, has jumped into the discussion of automated vehicles. Researchers around the world have conducted surveys and human subject experiments centered around trolley car scenarios [10, 11]. These studies attempt to crowdsource public opinion on what an autonomous vehicle should do in a no-win crash scenario when it is confined to the choice of hitting one entity over another.

Such a narrow framing of the problem contrasts with the way engineers actually program automated vehicles. The motion of a vehicle is not chosen at a single point in time by weighting known outcomes but rather evolves from a series of choices of the desired speed and path to take through an uncertain and ever-changing environment. Despite the lack of connection to the challenges faced by engineers developing motion planners, the plethora of recent papers and stories deploying no-win scenarios in this field has come to dominate the discussion of ethics for automated vehicles. It's common to find people who have the impression that engineers are (or should be) designing machines that solve these no-win moral dilemmas.

Early on in our work, we also interpreted the challenge of designing automated vehicles with ethics as a question of how to program ethics explicitly into computer algorithms [12]. To that end, in June 2015 we hosted an interdisciplinary workshop at the Center for Automotive Research at Stanford, including philosophers, engineers, lawyers and psychologists, titled "Towards Programming Ethics in Automated Vehicles." This workshop was a great success. Not because we came away with a clear answer on how to program ethics, but because one philosopher, Dr. Shannon Vallor of Santa Clara University, challenged us to think about the problem differently. In her opening talk, she stated that the problem we faced wasn't one of trying to program ethics into vehicles, but rather to find ways to program ethically. According to Vallor, the name of our workshop misplaced the focus, in that we should be focusing on developing processes and techniques that encourage engineers to program ethically. By using the language of values, we hope to put the focus back where it belongs and address the real ethical issues that arise with automated vehicles.

4 A Value-Centered Approach

So how do we put human values at the center of automated vehicle design? In one sense, the answer is strikingly obvious: we simply need to begin the conversation from the standpoint of human values. Values can form a common language across stakeholder groups, helping to identify potential value tensions up front and informing the search for resolutions. Admittedly, though, this is not a language that most people use in their day-to-day discussions of engineering problems or automated vehicles. Some tools and methods are therefore needed to facilitate this discussion.

We are far from the only ones to recognize the importance of human values in engineering design; some engineers routinely approach design from this perspective. Nor are we the only ones to highlight the need for methodologies that better connect human values and engineering. A variety of design processes with overlapping terminology such as "value", "values" and "worth" [13] have been developed to make some of these connections. Much of our inspiration comes from Value Sensitive Design (VSD), a general design methodology that formalizes the connection between human values and technology [14, 15]. As a methodology, VSD asks the designer to focus on a broad set of stakeholders that may be implicated by the designed technology (e.g. users, policy makers, the environment, the public), the values of those stakeholders, and the value tensions that may exist between different competing values. VSD is not restrictive to a formal process on achieving the integration of human values into the designed technology and, hence, is very open-ended in that respect.

We have been developing a structured approach to workshops that facilitate convening stakeholders, familiarizing them with the language of values and identifying values and value tensions from their input. Our approach endorses VSD's core considerations and may look very similar to VSD in specific applications [16]. In many ways our approach acts as a "wrapper" around VSD, constraining it for use in an engineering environment by providing a process and some specific tools for facilitating value-based conversations.

In this value-centered approach, we start by bringing together a wide-range of stakeholders as part of a stakeholder engagement session. When designing automated vehicles, such stakeholders could be executives and engineers from an automated vehicle company, policy makers, transportation officials, uniformed services or even local business owners and citizens. The broader the group of stakeholders, the wider the range of perspectives the designers will be able to consider. Bringing together this group of stakeholders helps to surface underlying values at stake in the design task. It also helps to identify value tensions and potential types of resolutions. The various perspectives can also help to frame the underlying problem in the design task. We have found that the key to an effective workshop is to base discussions around a scenario that is specific enough for people to understand, reveals the key value tensions and is accessible by stakeholders with different backgrounds [17].

After uncovering the value tensions in the design task, we can strategically approach the process of resolving these tensions. While there are many possible approaches to resolving value tensions, three that we have found to be particularly effective are:

(1) Prioritize – resolve the tension by choosing one value as the most important. Remaining values can then be addressed within the design space where this primary value is satisfied. For instance, in the case of truck platooning, safety can be chosen as the primary value. Once an appropriate level of safety has been set, the other values can be fulfilled to the greatest extent possible while maintaining this level of safety.
(2) Compromise – resolve the tension by balancing among competing objectives. In this approach, the goal is to partially fulfill the wishes for multiple values. Such an approach is a natural resolution to the conflicts in a routing algorithm, where neighborhood concerns could be balanced with travel time and traffic impacts in a cost function. When values can be appropriately quantified, this approach lends itself well to analysis of pareto optimality.
(3) Reframe – resolve the tension by changing the problem and dissolving the tension. One of our example scenarios involves tensions between a pedestrian and vehicle at a crosswalk. It is possible to envision road designs or future traffic control approaches where rights of way are handled entirely differently, making the original problem disappear [18]. Solutions of this type are often the most effective but can require significant change.

While workshops can be highly effective at identifying values, tensions and strategic approaches to resolving value tensions, the task of ensuring this information is incorporated into the design falls to the engineering team. Within the framework of VSD, this occurs during the technical implementation phase. To harness the information derived from the workshop, the technical implementation should provide a means for realizing the identified human values. Additionally, it is important to map the design parameters (such as the following distance from the platooning semi-truck example) to the associated values. This serves as clarification and a way to be explicit about what values will be captured in the system. Lastly, the technical implementation should enable treatment of the value conflicts. The strategies identified during the workshop provide an initial framing for how tensions or conflicts can be resolved. An important part of the technical implementation process is to determine whether or not this framing still seems appropriate with the greater level of specificity needed to make hard ethical choices. Even in the event that these approaches need to be revisited and revised, the fact that they were determined in the first place makes it easier for engineers to reengage stakeholders in a structured way.

Once a technology is implemented, we can analyze the system to determine if we captured the values appropriately. This entails revisiting the values and tensions identified during the stakeholder engagement. It may also include reconvening with some of the original stakeholders in the engagement session. This analysis component is a great opportunity for interdisciplinary discussion. Engineers can evaluate the design using simulation or even conduct user studies to gain feedback on the experience. Societal impacts can be estimated if appropriate models - such as traffic simulations or economic projections - are available. Ethicists can provide a rigorous critical analysis of the underlying values and justificatory approaches for resolving value tensions. Lawyers can provide a legal analysis given a particular implementation by conducting jury research or looking over case law.

A design process that is centered around human values creates common terminology for an interdisciplinary analysis that spans stakeholder groups. We believe a design process focused on human values and tensions will enable designers of automated vehicles to design technology that is societally accepted because potential value tensions have been identified and addressed ahead of time, thus aligning resulting automated vehicles with a robust set of value-centered justifications.

5 Conclusion

Engineers of automated vehicle technology make many small engineering decisions that can accumulate into systemic behavior that impacts society in a myriad of ways. It may not be obvious to all engineers which human values may be implicated in a design task, so we propose that a value-centered approach will help bridge the gap between engineering decisions and positive social impact. First, stakeholder engagement allows for a diverse and inclusive set of perspectives to enter the design process thus surfacing key values and value tensions. Secondly, designing the technology around those identified human values and tensions will help to bring about engineering implementations that help to resolve them earlier on in the design process. Finally, analysis based around human values allows for interdisciplinary discussion by using the values as common terminology. By advocating a human values-centered approach to designing automated vehicles, we believe society will be able to partake in the full benefits automated vehicles promise.

Acknowledgements. The authors would like to thank Daimler AG and the Ford Motor Company for their support of this research and for participating in our workshops.

References

1. Lopez, S.: LA Times. https://www.latimes.com/local/california/la-me-lopez-encino-waze-20180530-story.html (2018)
2. Foderaro, L.W.: New York Times. https://www.nytimes.com/2017/12/24/nyregion/traffic-apps-gps-neighborhoods.html (2017)
3. Borning, A., Muller, M.: Next steps for value sensitive design. In: Proceedings of the SIGCHI Conference on Human Factors in Computing Systems, pp. 1125–1134 (2012)
4. Bouton, M., Nakhaei, A., Fujimura, K., Kochenderfer, M.J.: Scalable decision making with sensor occlusions for autonomous driving. In: Proceedings of the IEEE International Conference on Robotics and Automation (2018)
5. Wongpiromsarn, T., Karaman, S., Frazzoli, E.: Synthesis of provably correct controllers for autonomous vehicles in urban environments. In: Proceedings of the IEEE Conference on Intelligent Transportation Systems, pp. 1168–1173 (2011)

6. Hart, P.E., Nilsson, N.J., Raphael, B.: A formal basis for the heuristic determination of minimum cost paths. IEEE Trans. Syst. Sci. Cybern. **4**(2), 100–107 (1968)
7. Lammert, M.P., Duran, A., Diez, J., Burton, K., Nicholson, A.: Effect of platooning on fuel consumption of class 8 vehicles over a range of speeds, following distances, and mass. SAE Int. J. Commercial Veh. **7**(2), 626–639 (2014)
8. Thomson, J.J.: Killing, letting die, and the trolley problem. Monist **59**(2), 204–217 (1976)
9. Greene, J.D.: Moral Tribes: Emotion, Reason, and the Gap Between Us and Them. Penguin, London (2014)
10. Bonnefon, J.-F., Shariff, A., Rahwan, I.: The social dilemma of autonomous vehicles. Science **352**(6293), 1573–1576 (2016)
11. Wächter, M.A., Faulhaber, A., Blind, F., Timm, S., Dittmer, A., Sütfeld, L.R., Stephan, A., Pipa, G., König, P.: Human decisions in moral dilemmas are largely described by utilitarianism: virtual car driving study provides guidelines for ADVs. ArXiv e-prints, 2017. arXiv: https://arxiv.org/abs/1706.07332 [cs.CY]
12. Gerdes, J.C., Thornton, S.M.: Implementable ethics for autonomous vehicles. In: Maurer, M., Gerdes, J.C., Lenz, B., Winner, H. (eds.) Autonomous Driving: Technical, Legal and Social Aspects, pp. 87–102. Springer, Berlin (2016)
13. Kujala, S., Väänänen-Vainio-Mattila, K.: Value of information systems and products: understanding the users' perspective and values. J. Inf. Technol. Theory Appl. **9**(4), 4 (2009)
14. Friedman, B., Kahn Jr., P.H.: Human values, ethics, and design. In: Jacko, J.A., Sears, A. (eds.) The Human-Computer Interaction Handbook, pp. 1177–1201. Lawrence Erlbaum Associates, Mahwah (2003)
15. Friedman, B., Kahn Jr., P.H., Borning, A.: Value sensitive design and information systems. In: Zhang, P., Galletta, D. (eds.) Human-Computer Interaction and Management Information Systems, vol. 5, pp. 348–372. M.E. Sharpe, Armonk (2006)
16. Thornton, S.M., Lewis, F.E., Zhang, V., Kochenderfer, M.J., Gerdes, J.C.: Value sensitive design for autonomous vehicle motion planning. In: Proceedings of the IEEE Intelligent Vehicles Symposium (2018)
17. Millar, J., Paz, D., Thornton, S.M., Parisi, C., Gerdes, J.C.: Design for human values – a framework for addressing ethical considerations in the engineering of automated vehicles (and other technologies), in preparation (2019)
18. National Association of City Transportation Officials: "Blueprint for Autonomous Urbanism," Module 1, Designing Cities Edition, Fall 2017

Four Perspectives on What Matters for the Ethics of Automated Vehicles

Geoff Keeling[1(✉)], Katherine Evans[2], Sarah M. Thornton[3], Giulio Mecacci[4], and Filippo Santoni de Sio[4]

[1] Leverhulme Centre for the Future of Intelligence, University of Cambridge, 16 Mill Lane, Cambridge CB2 1SB, UK
gk16226@bristol.ac.uk

[2] Sorbonne Universités and VeDeCom, 1, rue Victor Cousin, 75230 Paris cedex 5, France
katherine.evans@vedecom.fr

[3] Stanford University, 416 Escondido Mall, Building 550, Room 136, Stanford, CA 94305, USA
smthorn@alumni.stanford.edu

[4] Department of Values, Technology and Innovation, Section of Ethics and Philosophy of Technology, PO Box 5015, 2600 GA Delft, The Netherlands
{g.mecacci, f.santonidesio}@tudelft.nl

Abstract. The ethical discussion on automated vehicles (AVs) has for the most part focused on what morality requires in AV collisions which present moral dilemmas. This discussion has been challenged for its failure to address the various kinds of risk and uncertainty which we can expect to arise in AV collisions; and for overlooking certain morally relevant facts which are unique to the context of AVs. We take these criticisms as a starting point and outline four perspectives on what matters for the ethics of AVs: risk and uncertainty, value sensitive design, partiality towards passengers and meaningful human control.

Keywords: Autonomous vehicles · Robot ethics · Ethics of technology · Ethics of risk · Meaningful human control

1 Introduction

The ethics of automated vehicles (AVs) has received much attention in recent years. Some people have developed accounts of what morality requires in AV collisions which present moral dilemmas. These accounts are intended to reflect certain ethical commitments, such as the view that the minimization of harm is what matters in collisions; or the view that the justifiability of harm to the affected parties is most important [1–5]. Others have argued that the question of how AVs should allocate harm or risks of harm between road-users in collisions is a societal question, the solution to which at least in part depends on the preferences of road-users [6, 7]. Yet others have examined who is responsible for harm caused by AVs in collisions [8]. Parts of this discussion have been challenged for their failure to account for the various kinds of risk

G. Meyer and S. Beiker (Eds.): AUVSI 2019, LNMOB, pp. 49–60, 2019.
https://doi.org/10.1007/978-3-030-22933-7_6

and uncertainty which we can expect to arise in AV collisions; and for their failure to account for certain morally relevant considerations which are unique to the context of AV collisions, e.g. the special obligations which AV manufacturers have to passengers [7, 9, 10]. In this paper, we take these criticisms as a starting point, and outline four perspectives on what matters most for the ethics of AVs. In §2, Geoff Keeling argues that philosophers need to take as a starting point the different kinds of risk and uncertainty which we can expect to arise in AV collisions. In §3, Sarah Thornton examines how value sensitive design might be used to address the problem of anticipating pedestrians. In §4, Katherine Evans argues that prioritizing the safety of passengers is not as morally dubious as some people have made out; and in §5, Giulio Mecacci and Filippo Santoni de Sio argue that *meaningful human control* is the required standard of control for AVs to meet the appropriate level of safety and accountability.

2 Risk and Uncertainty

Geoff Keeling: Some people argue that AVs might encounter collision scenarios where the AV cannot avoid harming at least one person; and a choice is required about how to distribute harms or risks of harm between multiple people whose interests are in conflict [4, 5, 11, 12]. In response, several people have made *collision optimization* algorithms, which are sets of instructions for how AVs should distribute harms or risks of harm in collisions. Here are some examples: (1) Bonnefon et al. discuss a utilitarian algorithm, according to which AVs should select the action which causes the least harm [6]. (2) Leben defends a contractualist algorithm [1]. Roughly, contractualists care about the justifiability of harms to the affected parties as opposed to minimizing overall harm [5]. Leben's view is that AVs should distribute harm in accordance with the *maximin* decision rule: compare the worst-case scenario for each action and select the action with the best worst-case scenario. (3) Contissa et al. argue that AVs should be equipped with a personalized ethics setting, which enables the AV's passengers to determine the degree to which the AV prioritizes *their lives* over the lives of *other people* [3].

These algorithms have attracted some criticism. One problem is that, in different ways, they fail to account for the risk and uncertainty which we can expect to arise in AV collisions [7, 9, 10]. AVs have fallible sensors. From the AV's point of view, there are different ways the world might be, and the outcome of a collision depends on both the AV's action and the true state of the world. For example, the AV might be uncertain about the behaviors of pedestrians or it might be uncertain about morally relevant characteristics of the affected parties such as age and physical condition. It would be a stretch to say that existing algorithms are insensitive to considerations about risk and uncertainty *entirely*. For example, in arguing about the prospects for a contractualist algorithm, Leben and I are principally concerned with how AVs should distribute *risks* of harm between the affected parties in collisions [1, 5]. But we can sharpen this general criticism by articulating a particular kind of uncertainty which all these collision optimization algorithms overlook.

The collision algorithms described above all assume that the AV knows the number of affected parties in the collision. The utilitarian algorithm requires the AV to know how many people stand to be harmed conditional on each action available to the AV. Leben's contractualist algorithm assumes that the AV can calculate the survival probability of each affected party. And Contissa et al.'s algorithm assigns a utility function to each affected party, and then weights these utility functions in accordance with the passenger's preferences about the degree to which the AV prioritizes their lives over the lives of other people.

The assumption that the AV knows how many affected parties are involved in the collision is problematic for two reasons. The first is that, in the absence of vehicle-to-vehicle communication, the AV cannot be expected to know how many people are in each vehicle. So, defenders of collision optimization algorithms need to tell a plausible story about how the risks of harm to people inside other vehicles will be factored into their algorithms. It might be that statistical estimates of the number of persons in each vehicle are incorporated into collision algorithms. But there are ethical decisions to be made about the sorts of estimates used. The simplest approach is to take the AV's rational expectation of the number of people in each vehicle. This is the sum of each number of people who might be in the vehicle multiplied by the probability that the vehicle contains that number of people. If, for example, there is a 50% chance that a vehicle contains one person and a 50% chance that it contains two people, then the expected number of passengers is 1.5. This view has some problems. First, one implicit assumption in this view is that *underestimating* and *overestimating* the number of people in a vehicle are equally bad. It might be argued that the AV has reason to assign additional weight to the possibility that the vehicle contains a greater number of people. Second, once the appropriate degree of risk-aversion is determined, it remains an open question exactly how the risks of harm to, say, 1.5 (risk-weighted) expected people are balanced against risks harm to the AV's passengers and to pedestrians.

The second problem is that AVs use *classifier algorithms* to make predictions about what *kinds* of objects are in their environments. Many of these algorithms are probabilistic. So, for a given object, the AV might be 50% certain that the object is a person and 50% certain that it is a tree. Defenders of collision optimization algorithms owe an account of the conditions under which the AV ought to behave *as if* an object is a pedestrian, given that we can expect all AVs to be uncertain to some degree about what kinds of objects are in their environments. Two plausible views are that the AV should behave as if an object is a person just in case the AV's credence in that object being a person exceeds some numerical threshold; and that the AV should behave as if the object is a person if that object is more likely to be a person than it is likely to be something else. What I take to be the most plausible view is that the AV's degree of aversion to colliding with an object should be a continuous strictly increasing function of its credence in that object being a person. But this view cannot easily be squared with existing collision optimization algorithms. Perhaps this does not matter. My view is that the role of philosophers is not to design collision algorithms, but to make precise the different kinds of risk and uncertainty in AV collisions and to discuss with engineers what morality requires in decisions involving those kinds of uncertainty. The task of *designing* collision optimization algorithms is better left to engineers.

3 Value Sensitive Design for Motion Planning

Sarah M. Thornton: At the core of motion planning algorithms are the variables that parameterize them. These values can have ethical implications. If engineers can connect those parameters to human values, such as mobility, safety, and legality, then stakeholders can have a better understanding of these ethical implications. Value sensitive design (VSD) is a tool that formalizes the connection of human values to engineering specifications through a process of discovery and iteration.

The VSD methodology has three phases: conceptual, technical, and empirical [13, 14]. The conceptual phase involves identifying the values encompassed by the designed technology and determining the stakeholders of the technology. A feature of VSD holds that some technological implementations are better suited to uphold certain values than others. For the technical phase, the technical solutions most in line with the identified values are used to develop the technology. Finally, the empirical phase allows for quantitative and qualitative analyses of the developed design. This period allows for inspection of how successfully the designed technology meets the conceptualization. Throughout the design development, the designer iterates over the various phases until all three align.

Conceptualization: We focus on a case-study to keep things simple. Figure 1 depicts a two-lane roadway with a single, dashed yellow line and a marked pedestrian crosswalk. In front of the crosswalk is an illegally parked van. From the perspective of the AV approaching the crosswalk, the crosswalk is partially occluded because of the obstructing van. The design task is to develop a speed control algorithm along the given obstacle-free path such that the AV safely navigates through the scenario.

The stakeholders involved with the scenario and design task are identified in addition to the human values. For this first iteration, we consider the stakeholders to be the AV, occupants in the AV, the pedestrian potentially crossing the street, the authority of traffic laws, and the obstructing vehicle parked on the road. There are many more stakeholders to consider in future iterations, such as bicyclists or bystanders. Traffic scenarios, in general, relate to balancing the human values of safety, legality, and mobility. By considering the stakeholders, we uncover more values at stake, such as care and respect for others, fairness and reciprocity, respect for authority, trust and transparency, and individual autonomy.

The California Vehicle Code §21950 suggests following the law and driving safely strongly correlate. For this iteration, we assume safety and legality to be the same engineering specification for this scenario, which corresponds to the human values of care and respect for others, respect for authority, fairness and reciprocity, safety, and legality. It can be captured by vehicle speed (v_t), distance to crosswalk (d_t), and whether or not a pedestrian is crossing the street (c_t). The metric of time efficiency captures the human values of mobility and individual autonomy, and efficiency relates to the speed of the vehicle (v_t). Smooth driving affects occupant comfort and interjects trust and transparency between the stakeholders. Hence, smoothness can be captured through the change in vehicle speed, which is equivalent to the acceleration command (a_t) multiplied by the change in time (Δt).

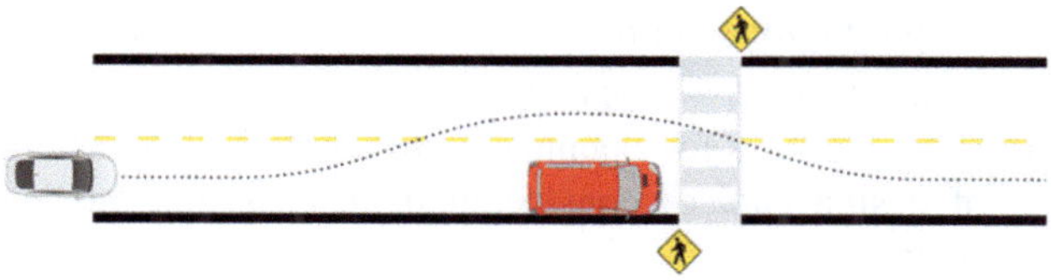

Fig. 1. Experimental scenario of occluded pedestrian crosswalk.

Technical Implementation: For this design task, in order to obtain an offline policy to inspect and verify before putting on an AV, a closed-loop planning approach is chosen and the problem is constructed as a partially observable Markov decision process (POMDP) [15]. Throughout the design of the POMDP, every design choice is connected back to values from the conceptualization phase in order to justify the engineering and explicitly record the embedding of said values.

We describe only the reward function $R(s, a)$ of the POMDP, which defines the immediate reward for every state and action. Although there are other components in the POMDP design that can impact the accounting of human values, the reward function most clearly connects the conceptualization values to the technical implementation. The negative reward (or penalty) term for safety and legality is:

$$R_{\text{safe}}(s, a) = -\left(\zeta \frac{v_t^2}{d_t + \epsilon} + \eta \mathbf{1}(d_t = 0)\right) \mathbf{1}(c_t),$$

where $\epsilon > 0$ is a buffer to soften the constraint, $\zeta > 0$ is a weight on the penalty incurred by driving quickly as the vehicle gets closer to the crosswalk, $\eta > 0$ is a terminal penalty independent of velocity to encourage the vehicle to stop when the pedestrian is crossing, and $\mathbf{1}(\cdot)$ is a function that evaluates to 1 if the Boolean logic is true and 0 if false. The reward term for efficiency and mobility is:

$$R_{\text{eff}}(s, a) = \lambda v_t \mathbf{1}(\neg c_t),$$

where $\lambda > 0$ is a reward weight to encourage higher speed when the pedestrian is not crossing. To achieve smoothness for occupant comfort, the objective is realized through a penalty term on the change in velocity:

$$R_{\text{smooth}}(s, a) = -\xi(v_t - v_{t+1})^2 = -\xi(a_t \Delta t)^2,$$

where $\xi > 0$ is the weight penalizing large changes in velocity. The total reward for a state-action pair is the sum of $R_{\text{safe}}(s, a)$, $R_{\text{eff}}(s, a)$, and $R_{\text{smooth}}(s, a)$. In order to solve the POMDP, the method of QMDP is used to approximate an optimal solution [15].

Empirical Analysis: The analysis comes from experiments on an automated Ford Fusion. The VSD process is compared to a baseline to determine its impact on a control speed algorithm. The baseline is a deterministic proportional speed control. Once the pedestrian is detected, a constant deceleration is commanded based on the current vehicle velocity and distance to the crosswalk. When no pedestrian is detected, then the vehicle resumes a proportional cruise control with gain (k_p) and known desired velocity (v_{des}) as in 'if c_t, $a_t = -v_t^2/2d_t$, else $a_t = k_p(v_{des} - v_t)$.' The baseline is intentionally simple as it allows for examination of the design characteristics due to the limited number of design choices considered in this baseline implementation. The results in Fig. 2 have many properties to improve, such as the smoothness of the acceleration commands, but it does depict that the design decision to model the problem as a POMDP and solve for an offline policy helped with investigating and balancing some of the value tensions in this design task. This is evidenced by the vehicle approaching cautiously to the crosswalk in anticipation of a pedestrian potentially appearing from behind the occluding vehicle.

This section demonstrates the formal connection of human values into the design of a speed control algorithm through the conceptualization and technical implementation phases. The empirical analysis phase helps identify areas of improvement for subsequent iterations. For more details, please see [16].

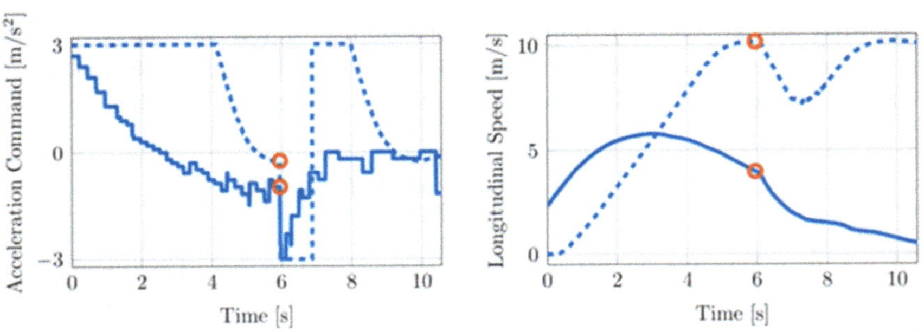

Fig. 2. Plots of the commanded acceleration (left) and measured speed (right) for both the baseline (dashed) and POMDP (solid) implementations synchronized to the point in time of detecting the pedestrian (red circle).

4 Why Not Put the Passenger First?

Katherine Evans: On a fateful day in October 2016, Mercedes executive Christophe von Hugo expressed in an interview with *Car and Driver* a statement from which there would be no turning back: 'Save the one in the car. If all you know for sure is that one death can be prevented, then that's your first priority' [17]. This humble maxim, capturing the proscription that 'an A.I. should prioritize the lives it has the most direct control of', received such a cacophony of press blowback that it forced von Hugo to reverse his position only days later. Nevertheless, philosophical inquiry sometimes requires investigating what is unpopular, and thus, the question must be asked: why *not* put the passenger first?

The tragedy of the 'passenger first' policy is its imagined connection to *moral exclusivism*. By protecting the passenger *at all costs*, the AV will cause disproportionate harm to others; and the iniquity of killing a pedestrian to avoid passenger whiplash is hardly a moral mystery. But it is not moral exclusivism with which we must contend, but a form of *partiality* regarding the passenger's welfare. In effect, 'passenger first' is a scalar concept, the most extreme form of which is moral exclusivism where the vehicle is prepared to sacrifice any vulnerable road user as soon as the passenger is even the slightest bit at risk. This position is untenable. But there are two further positions that capture the concept, both of which must be discounted if 'passenger first' is to be rejected.

The first is what one might call a *firm moral preference* for the passenger. Under this conception, the harm that the vehicle inflicts on other road-users is *collateral harm* resulting from the vehicle's protection of its passenger's life. This moral preference for the passenger is more proportionate than moral exclusivism since it is the act of *saving the passenger's life* that prompts the vehicle to collaterally harm other road-users. However, like moral exclusivism, this view is not above scrutiny. Firstly, the 'moral preference' position still hands the passenger an *a priori* assurance that the passenger will never suffer deliberate lethal harm at the hands of his or her AV. This may still fall short of the demands of justice and equity, as the passenger can still matter disproportionally more than any other road user in the vehicle's environment. Moreover, the moral preference position is silent on the proportional treatment of these *other* road users and may therefore cause unnecessary collateral harm to those it sacrifices in the pursuit of the safety of its passenger.

There is a weaker position on offer: *minimal passenger partiality*. Under this conception, the harm inflicted on other road-users is *predictable collateral harm* resulting from the vehicle's protection of its passenger. The passenger's welfare is still prioritized, but in some *proportional relation* to the welfare of other road users. The AV's directive is to save its passenger's life *and* minimize the predicted harm it will inflict on other road-users as a consequence. The proportional relation between passenger and other road-users can vary in form. It can yield decisional policies that still afford an existential 'trump' card to the passenger, following the doctrine of double effect; or, it can follow along more utilitarian lines, in which the passenger may be sacrificed if the sacrifice of a greater number of lives is required to avoid the passenger's death. In all cases however, minimal passenger partiality requires a maximum degree of proportionality, in respect to both the passenger and the surrounding road users. Minimal passenger partiality is the most agreeable definition of 'passenger first'. Firstly, in the case of lethal unavoidable collisions, harm to more than one party is generally foreseeable, even if certain details such as probability of survival are difficult to ascertain. In this way, any decisional policy that treats this harm as a purely unpredictable collateral harm is morally irresponsible. Finally, in terms of equity and justice, minimal passenger partiality affords the most proportionate response to the claims of the different road users implicated in the vehicle's decision context.

However, for some, this minimal passenger partiality may still prove to be too 'passenger-centric'. Even this minimal form of 'passenger first' affords a certain 'priority' to the passenger, and therefore is guilty of unequal treatment. If minimal passenger partiality is too biased for comfort, perhaps a deeper question must be asked: what is the morally optimal *teleology* of AVs? In other words, what *ought* to be the larger decisional goal of AVs, to protect the passenger first, or to protect all road users equally? At first blush, the moral nose seems to advocate for an impartial decisional policy, one that aligns with liberal values, rationalist moral theory [18], and the so-called utilitarian preferences of certain populations within empirical studies [6]. But does impartial decision-making best reflect the realities of the modern traffic environment or *current* user expectations? One might cautiously defend the contrary for three reasons. Firstly, AVs will arrive incrementally. For society to benefit from the advantages of AVs, users must *want* to step into these vehicles. Minimally, a form of passenger partiality may act as an incentive to use this emergent technology. Of course, this is not to say that a more impartial decisional policy could not be implemented once human driving is a thing of the past. Secondly, the question of initial user expectations must be asked. Whether in a robot taxi, or a privately-owned vehicle, the passengers of the initial stages of AV implementation will be human drivers; equipped with human instincts, behaviors and traditional conceptions of property rights and service agreements. Will they expect their AVs to prioritize their safety to some degree? Whatever the answer, it is conceptually separate from what is impartially good or right. Finally, imbedded in the polemics surrounding AVs is the human reaction to a new-found loss of *decisional autonomy*. As the world of AI expands into more and more human decisions, so too do the risks of technological paternalism and decisional impotence [19]. It is clear that technologists have an ethical duty to align their systems with user expectations, *a fortiori* in those technologies that could prove lethal to the user. Perhaps this ethical duty does not overpower the civic duty we have to avoid harm where possible. Or, perhaps current user expectations are less atomistic and more impartial. Nevertheless, these are the questions that must be answered before one can categorically refuse to put the 'passenger first'.

5 Meaningful Human Control Over Autonomous Driving Systems

Giulio Mecacci and Filippo Santoni de Sio: Many see high levels of automation as the solution to several problems, in traffic and elsewhere. However, recent literature mentions numerous reasons why we should make sure automated vehicles ultimately remain under the control of some responsible human agent. Some authors stress how genuine human involvement in control tasks might make some operations safer, insofar as human cognitive capacities may best compensate for some of the current limitations of artificial intelligence [20–22]. Many have also pointed out that automation might create undesirable "responsibility gaps" [23]. Since the distribution of control tasks within the human-machine system tends to be obscure, its actions might be hard to clearly account for. This is particularly problematic in fields where stakes are high, such as automated warfare [24, 25] and automated driving [26, 27].

In particular, in the political debate over autonomous weapon systems, there has been a convergence around the idea that "meaningful forms of human control" ("MHC" henceforth) over automated systems should at all times be maintained, i.e. there should always be a human being who is ultimately responsible for the operation of the military [28, 29]. This form of control, though intuitively desirable, might just not be possible. The main reason seems to be that automation is inherently incompatible with human control. More automation simply entails less human control. In fact, the tasks that we decide to automate are tasks that machines are supposed to do better or more efficiently than we do.

Schwarz [30], in an influential blog post on Humanitarian Policy and Law, recently mentioned three reasons why MHC over autonomous weapon systems might be (close to) impossible. Those reasons, we argue, seem to equally apply to war drones and automated vehicles. Both of them involve intelligent systems and automated decision making.

The first reason concerns the cognitive limitations that emerge in the human-machine interaction. Humans tend to "lay back" and display superficial reasoning and reflection when interacting with highly automated systems. This drastically diminishes the quality and quantity of a human controller's contribution and commitment to the overall decision-making process. Another reason why human control might be hard to achieve is that humans cannot access and effectively make use of the amount of information that an automated intelligent system typically processes in its functioning. That is after all the very reason we let computers and assistive technologies take over some of our tasks. However, meaningful involvement in decisions is hard if we cannot see the grander scheme of things [31]. A third issue would regard temporal limitations. Due to both cognitive and physical limitations, humans are very slow when it comes to timely and resolutely intervening to steer a last second decision that a device might have suggested or plainly made.

If humans are unable to expediently understand, question and act upon decisions and actions that are mediated by automated systems, the prospect of meaningful control seems practically excluded, and with it, one could argue, go some degree of safety and the possibility for reasonable moral and practical accountability. Or does it? We believe that the three concerns presented, among others, are all sound. However, we argue that those human shortcomings only hold so long as one endorses a certain notion of MHC.

MHC, in our perspective, should not be intended as *more* control. Surely not more of the same kind. If we stick to a classic notion of control [32], a controller (be it human or otherwise) is in control of a system so long as they are directly, operationally so. In other words, insofar as the system causally responds to the action of a (human) controller upon the system's operation. Automation in general, of which assistive driving systems are a prime example, inherently makes this kind of control hard to achieve and maintain. It introduces variables between human behavior and the resultant decisions, thereby making them less predictable. Moreover, it makes the role of the human controller (in the case of automated vehicles: the driver) less and less relevant.

Yet MHC, we believe, is not about the amount of that kind of control, which is indeed doomed to decrease. Rather, it is a qualitatively different one that carves the nature of control at different joints. It is not about the responsiveness of the system to the *actions* of the controller, but rather about its responsiveness to the relevant human *reasons*, including those of the drivers, as well as of the designers, regulators, other humans interacting with the system etc. The concept of MHC can be understood by looking at the relation human persons have, for instance, to a horse. The horse metaphor ("the H-metaphor") is not new. It first appears in Flemisch et al. [33], although the intuition was already present earlier in Michon [34], when he framed *strategic* control as a higher order level of control in road users' driving tasks. MHC is the kind of control that gives us criteria suitable to measure how well the horse will behave in terms of the expectations we have for its behavior. This in turn depends, as we argue, not only on the horse's responsiveness to the orders of the horse rider, but also, on its responsiveness to the training, including the capacity to respond to the right elements of the environment, to ignore misleading cues etc. Indeed, it is when we as horse riders are confident that the horse is sufficiently responsive to this latter set of reasons, that we can loosen the reins (operational control), while being confident that it will behave exactly the way we would want it to (MHC). This notion of control is fascinating because it also allows an expansion of the range of possible controllers to those individuals that are not directly in contact with the controlled systems, but are at some point involved in its design, deployment or regulation.

Now, from horses – to horsepower, this is what we suggest MHC should look like in relation to automated driving systems. In our work[1], following de Sio and van den Hoven [27], we isolated two criteria for MHC over automated driving systems, termed 'tracking' and 'tracing', and we currently aim to operationalize them into design guidelines. We have applied our theory of MHC to case studies ranging from truck platooning [35] to dual mode vehicles [36]. For those use cases, we used our criteria of MHC as benchmarks to (i) assess the degree to which a vehicle is under MHC in certain circumstances (e.g. accidents) as well as to (ii) assess its principled *controllability* and (iii) suggest design guidelines to possibly improve it. Future research will aim at establishing increasingly stringent criteria for a meaningful control interaction between controllers and controlled systems, not only under a technical and infrastructural design perspective, but also looking at limits and potentials of human behavioral capacities [37]. We ultimately aim to provide recommendations for testing, standardization and driving licensing procedures which promote sufficient levels of meaningful human control.

6 Conclusion

In this paper, we offered four perspectives on what matters for the ethics of automated vehicles. First, Geoff Keeling argued that moral philosophers addressing the question of what morality requires in AV collisions ought to take as a starting point the different kinds of risk and uncertainty which we can expect to arise in these collisions. Second, Sarah M. Thornton argued that the variables used in motion planning algorithms have ethical implications; and that value-sensitive design provides a plausible methodological framework for designing these algorithms in accordance with morally relevant considerations. Third, Katherine Evans argued that prioritizing the welfare of AV passengers in collisions is not as morally dubious as people often claim in the media. Fourth, Giulio Mecacci and Filippo Santoni de Sio argued that meaningful human control is the required standard of control for AVs to meet the appropriate level of safety and accountability.

[1] This work is part of the research project Meaningful Human Control over Automated Driving Systems with project number MVI.16.044, which is (partly) financed by the Netherlands Organisation for Scientific Research (NWO).

References

1. Leben, D.: A Rawlsian algorithm for autonomous vehicles. Ethics Inf. Technol. **19**(2), 107–115 (2017)
2. de Sio, F.S.: Killing by autonomous vehicles and the legal doctrine of necessity. Ethical Theor. Moral Pract. **20**(2), 411–429 (2017)
3. Contissa, G., Lagioia, F., Sartor, G.: The ethical knob: ethically-customisable automated vehicles and the law. Artif. Intell. Law **25**(3), 365–378 (2017)
4. Keeling, G.: Legal necessity, pareto efficiency and justified killing in autonomous vehicle collisions. Ethical Theor. Moral Pract. **21**(2), 413–427 (2018)
5. Keeling, G.: Against Leben's Rawlsian collision algorithm for autonomous vehicles. In: Müller, V. (eds.) Philosophy and Theory of Artificial Intelligence (PT-AI 2017). Studies in Applied Philosophy, Epistemology and Rational Ethics, vol. 44. Springer, Cham (2017)
6. Bonnefon, J., Shariff, A., Rahwan, I.: The social dilemma of autonomous vehicles. Science **365**(6293), 1573–1576 (2016)
7. Himmelreich, J.: Never mind the trolley: The ethics of autonomous vehicles in mundane situations. Ethical Theor. Moral Pract. **21**(3), 669–684 (2018)
8. Hevelke, A., Nida-Rümelin, J.: Responsibility for crashes of autonomous vehicles: an ethical analysis. Sci. Eng. Ethics **21**(3), 619–630 (2015)
9. Goodall, N.: Away from trolley problems and towards risk-management. Appl. Artif. Intell. **30**(8), 810–821 (2016)
10. Nyholm, S., Smids, J.: The ethics of accident-algorithms for self-driving cars: an applied trolley problem? Ethical Theor. Moral Pract. **19**(5), 1275–1289 (2016)
11. Goodall, N.: Ethical decision making during automated vehicle crashes. Transp. Res. Rec. **2424**(1), 58–65 (2014)
12. Goodall, N.: Machine ethics and automated vehicles. In: Meyer, G., Beiker, S. (eds.) Road Vehicle Automation. Lecture Notes in Mobility. Springer, Cham (2014)
13. Friedman, B., Kahn Jr., P.H.: Human values, ethics, and design. In: Jacko, J.A., Sears, A. (eds.) The Human-Computer Interaction Handbook, pp. 1177–1201. Lawrence Erlbaum Associates, Mahwah (2003)
14. Friedman, B., Kahn Jr., P.H., Borning, A.: Value sensitive design and information systems. In: Zhang, P., Galletta, D. (eds.) Human-Computer Interaction and Management Information Systems, vol. 5, pp. 348–372. M.E. Sharpe, Armonk, NY (2006)
15. Kochenderfer, M.J.: Decision Making Under Uncertainty: Theory and Application. MIT Press, Cambridge (2015)
16. Thornton, S.M.: Autonomous vehicle motion planning with ethical considerations. PhD thesis. Stanford University (2018)
17. Morris, D.Z.: Mercedes-Benz's self-driving cars would choose passenger lives over bystanders. Fortune (2016). http://fortune.com/2016/10/15/mercedes-self-driving-car-ethics/
18. Ogienr, R.: L'éthique aujourd'hui. Maximalistes et minimalistes, pp. 144–152. Gallimard, Paris (2015)
19. Millar, J.: Technology as a moral proxy: autonomy and paternalism by design. IEEE Technol. Soc. Mag. **34**(2), 47–55 (2015)
20. Brooks, R.: The seven deadly sins of AI predictions. MIT Technol. Rev. (2017). https://www.technologyreview.com/s/609048/the-seven-deadly-sins-of-ai-predictions/. Accessed 17 Jan 2019
21. Selman, B., Brooks, R., Dean, T., Horvitz, E., Mitchell, T., Nilsson, N.: Challenge problems for artificial intelligence. In: Proceedings of AAAI-96, Thirteenth National Conference on Artificial Intelligence, pp. 1340–1345 (1996)

22. van Gerven, M.: Computational foundations of natural intelligence. Front. Comput. Neurosci. **11**, 112 (2017)
23. Matthias, A.: The responsibility gap: ascribing responsibility for the actions of learning automata. Ethics Inf. Technol. **6**(3), 175–183 (2004)
24. Di Nucci, E., de Sio, F.S.: Drones and responsibility: mapping the field. Routledge (2016)
25. Horowitz, M.C., Scharre, P.: Meaningful Human Control in Weapons Systems: A Primer (2015)
26. Nyholm, S.: Attributing agency to automated systems: reflections on human-robot collaborations and responsibility-loci. Sci. Eng. Ethics **24**(4), 1201–1219 (2018)
27. de Sio, F.S., van den Hoven, J.: Meaningful human control over autonomous systems: a philosophical account. Front. Robot. AI **5**, 15 (2018)
28. Scharre, P.: Army of None: Autonomous Weapons and the Future of War. W. W. Norton, New York (2018)
29. Moyes, R.: Key Elements of Meaningful Human Control. Article 36 (2016)
30. Schwarz, E.: The (im)possibility of Meaningful Human Control for Lethal Autonomous Weapon Systems (2018). https://blogs.icrc.org/law-and-policy/2018/08/29/im-possibility-meaningful-human-control-lethal-autonomous-weapon-systems/
31. Mittelstadt, B.D., Allo, P., Taddeo, M., Wachter, S., Floridi, L.: The ethics of algorithms: mapping the debate. Big Data Soc. **3**(2), 1–21 (2016)
32. Bennett, S.: A brief history of automatic control. IEEE Control Syst. Soc. **16**, 17–25 (1996)
33. Flemisch, F.O., Adams, C.A., Conway, S.R., Goodrich, K.H., Palmer, M.T., Schutte, P.C.: The H-Metaphor as a Guideline for Vehicle Automation and Interaction, January 2003
34. Michon, J.A.: Human Behavior and Traffic Safety. Springer, Boston (1985)
35. Calvert, S.C., Mecacci, G., Heikoop, D.D., de Sio, F.S.: Full platoon control in truck platooning: a meaningful human control perspective. In: 21st International Conference on Intelligent Transportation Systems (ITSC), pp. 3320–3326 (2018)
36. Mecacci, G., de Sio, F.S.: Meaningful Human Control, Practical Reasoning and Dual-Mode Vehicles (2019, under review)
37. Heikoop, D., Hagenzieker, M., Mecacci, G., Calvert, S., de Sio, F.S., van Arem, B.: Human behaviour with automated driving systems: a quantitative framework for meaningful human control (2018, under review)

Automated Vehicles and Vulnerable Road Users: Envisioning a Healthy, Safe and Equitable Future

Justin M. Owens[1(✉)], Laura Sandt[2], Azra Habibovic[3], Sarah Rebolloso McCullough[4], Ryan Snyder[5], Robert Wall Emerson[6], Pravin Varaiya[7], Tabitha Combs[8], Fred Feng[9], Mohammed Yousuf[10], and Bernard Soriano[11]

[1] Virginia Tech Transportation Institute, 3500 Transportation Research Plaza, Blacksburg, VA 24061, USA
jowens@vtti.vt.edu

[2] University of North Carolina Highway Safety Research Center, 730 Martin Luther King Jr. Blvd, Suite 300, Chapel Hill, NC 27599-3430, USA
sandt@hsrc.unc.edu

[3] RISE Research Institutes of Sweden, Lindholmspiren 3A, 417 56 Gothenburg, Sweden
azra.habibovic@ri.se

[4] University of California, Davis Feminist Research Institute, One Shields Drive, Davis, CA 95616, USA
smcc@ucdavis.edu

[5] Transpo Group, 10501 Wilshire Blvd. #1910, Los Angeles, CA 90024, USA
ryan.snyder@transpogroup.com

[6] Department of Blindness and Low Vision Studies, Western Michigan University, Kalamazoo, MI 49008-5218, USA
robert.wall@wmich.edu

[7] Department of Electrical Engineering and Computer Sciences, University of California Berkeley, Berkeley, CA 94720, USA
varaiya@berkeley.edu

[8] Department of City and Regional Planning, University of North Carolina, New East 213, Chapel Hill, NC 27599-3140, USA
tacombs@live.unc.edu

[9] University of Michigan Transportation Research Institute, 4901 Evergreen Rd, 2320 HPEC, Dearborn, MI 48128, USA
fredfeng@umich.edu

[10] Turner-Fairbank Highway Research Center, FHWA, 6300 Georgetown Pike, Mclean, VA 22101, USA
mohammed.yousuf@dot.gov

[11] CA Department of Motor Vehicles, 2415 1st Ave, Sacramento, CA 95818, USA
bernard.soriano@dmv.ca.gov

Abstract. This chapter provides an overview and recap of the AVS 2018 Breakout Session #8, *AVs & Vulnerable Road Users: Envisioning a Healthy, Safe, and Equitable Future*, including summaries of research presentations,

G. Meyer and S. Beiker (Eds.): AUVSI 2019, LNMOB, pp. 61–71, 2019.
https://doi.org/10.1007/978-3-030-22933-7_7

perspectives on equity from leading experts in the field, and lessons learned through discussion among panelists and the session audience. The session identified a range of necessary actions and research needs such as engaging stakeholders at all levels from community to OEMs and governments to identify and solve problems before they become evident, recognizing the tradeoff between safety and access to transportation, and learning from the history of how transportation has affected (SES) communities.

Keywords: Equity · Automation · Vulnerable road user · Pedestrian · Cyclist · SES · Disability · Mobility · Accessible transportation · Travelers with disabilities

1 Introduction

The history of vehicle transportation is a story of both tremendous ingenuity and rapid advancement alongside a wholesale shift in how cities were designed and neighborhoods connected – or destroyed [1] – in service of a rapidly changing vision of mobility. In retrospect, some of the ideas of transportation progress during the mid-20th century, while primarily designed to improve access to freeways and reduce congestions, served to restrict the mobility and economic growth of some members of the population of the US, including those in low-socio-economic-status (SES) areas and people of color [2, 3]. Many of these restrictions directly impacted communities' walkability and mobility, both by increasing reliance on automobiles and decreasing the amount of infrastructure devoted to safe walking.

The goal of this breakout session was to expand the discussion surrounding the interactions among automated vehicles (AVs) and vulnerable road users (VRUs). In previous years, meetings of this session have focused on the technological and human factors considerations of roadway interactions between AVs and VRUs, specifically pedestrians and cyclists [e.g., 4]. While these considerations remain important and timely, and this session devoted several talks to discussion of the latest research surrounding them, we felt it was critical this year to go beyond physical interactions to consider the larger realm of opportunities and potential consequences that AVs may bring to the VRU landscape. Relatedly, we wished to expand the definition of VRU beyond pedestrians and cyclists to include members of society made vulnerable by their SES, disability, and/or other personal characteristics.

By expanding the conversation beyond purely physical pedestrian/cyclist/AV interactions to the broader mingling of equity and automation, we aim to begin the conversation surrounding an impending shift in our transportation landscape early on, enabling learning from some of the prior century's mistakes and ensuring that the benefits of AV technology are equally available for all and the risks are fairly spread among the population.

To accomplish this, the breakout session was structured to enable attendees to accomplish three distinct goals:

(1) Learn about the latest advances in technology designed to facilitate AV/VRU interaction

(2) Hear new and varied perspectives about the importance of equity, planning and design in ensuring safe and fair mobility in future transportation systems
(3) Engage with researchers, experts, and safety advocates to exchange ideas and identify research needs.

After a session introduction and overview, the first portion of the session was devoted to a series of four talks that presented recent research on ways that AVs and/or drivers currently interact with VRUs including seniors, pedestrians and cyclists, as well as an overview of the equity-related challenges that will be faced as we transition to AVs. The session then transitioned to a panel discussion featuring experts in a range of disciplines touching on equity and transportation, followed by interactive discussion with the audience. The session lasted for a single afternoon from 1:30–5:30. There were a total of 32 attendees, not including speakers and panelists, which represented a significant increase over session attendance in 2017. This chapter will be broken into three sections corresponding with the goals above.

2 Advances in Technology and High-Level Considerations

During the first section of the session, several speakers presented updates on considerations and on-going research related to technological and design characteristics of AVs that could mediate their interactions with VRUs.

Automation for Increased Accessibility and Inclusion
Azra Habibovic, Senior Researcher, RISE Research Institutes of Sweden
The mobility of children, elderly and people with disabilities is often limited, which may make them feel dependent on others and/or socially excluded. This is mainly due to poor street design and traffic complexity, public transportation that is not easily accessible, difficulty of parking close to destinations, complex interfaces that limit use of new technologies, and expensive vehicle adaptations. A recent study in Sweden [5] characterized a range of challenges that blind people face when using mobility services including difficulty planning, lack of information, and dealing with abrupt maneuvers.

The current development of automated vehicles and mobility services often does not take these and other considerations facing these vulnerable populations into consideration. For example, automated driving systems are frequently installed in existing vehicle models that are not designed with elderly and disabled people in mind. Similarly, entering and using seat belts in fully automated shuttles can require human support. On-demand services often pick up and drop off passengers in the middle of the street, exposing them to multiple risks. Services addressing the “first and last mile” run along predefined routes, and accessing them often requires walking and navigating in traffic; it remains an open question how vulnerable populations can safely interact with automated vehicles when encountering them in traffic.

To achieve improved mobility for *all* people, future developments should adopt universal design [6], which emphasizes that solutions need be usable by all people, to the greatest extent possible, without the need for adaptation or specialized design. Adopting universal design early in the development process is crucial to the successful use of automated technology by vulnerable populations, as is establishing necessary collaborations between the public and private stakeholders. This will help developing integrated solutions that address challenges along the entire trip: before, during, and after traveling in a vehicle.

Making Intersections Safer with I2X Communication
Pravin Varaiya, UC Berkeley Dept of Electrical Engineering & Computer Sciences
Intersections are very dangerous. Forty percent of all crashes, 50% of serious collisions, and 20% of fatalities occur in intersections. Bay Area fatalities increased 43% between 2010 and 2016 to reach 455 killed, of which, in San Francisco, 62% were cyclists or pedestrians. Intersections are challenging because of complex interactions among pedestrians, bicycles and vehicles, absence of lane markings to guide vehicles, split phases that prevent determining who has the right of way, obstructions from stopped vehicles, and illegal movements. Improving intersection safety is urgent.

AV manufacturers offer a radical path to safety, with claims that AVs can prevent 94% of all crashes involving human error. However, current AV technology is challenged by intersections; 58 of 66 (88%) AV crashes reported to the DMV occurred in intersections. Even if AVs eventually reach target levels of safety, it will be several decades before they are widely deployed. Meanwhile, pedestrians and cyclists face high risk of injury and death.

Our research seeks to remove one important cause of intersection accidents: drivers, pedestrians and cyclists make mistakes because they lack sufficient information about the movement of others as they proceed through an intersection. There is spatial and temporal uncertainty. Spatial uncertainty arises from the fact that the traffic signal we see as we approach the intersection does not announce who else entering from the other approaches has the right of way. This missing information can be supplied by an 'intelligent intersection'. It reports the traffic signal from all approaches; predicts when the signal phase will change; announces which blind spots are occupied using sensor data; and predicts red light violations before they occur. The intelligent intersection broadcasts this information via radio that can be received by everyone in the intersection with a smartphone or Bluetooth device.

We have designed the software and hardware that comprises the 'intelligence' for any signalized intersection. We can assess the resulting safety improvements. Upgrading an intersection will cost between $20 K and $90 K, depending on the size of the intersection and how many detectors it already has. Vision Zero investments may improve safety, but they can be expensive and reduce mobility. Automated vehicles may in 15 years yield safety benefits. The intelligent intersection complements both approaches because the information it provides serves as an additional safety buffer for pedestrians, cyclists and drivers.

Driver-Bicyclist Interactions Using Naturalistic Driving and Bicycling Data
Fred Feng, University of Michigan-Dearborn
As a more complete understanding of how drivers and VRUs interact can inform the development of AV technology, Dr. Fred Feng provided an overview of three studies examining how drivers interact with bicyclists in the real world. These included a study of how drivers overtake bicyclists [7] as well as ongoing research studying corner cases and driver-bicyclist interactions using naturalistic driving and naturalistic bicycling data. Methodologies include naturalistic studies featuring both vehicle and bicycle

instrumentations that provide continuous, detailed, and quantitative data on how drivers and bicyclists behave and interact with each other for their everyday trips on real-world roadways. The quantitative results from this work could be potentially used to develop, test, and benchmark automated vehicle technologies on how to interact with bicyclists safely and efficiently.

Hidden Equity Challenges of an AV-Dominant Transportation System: Tradeoffs Between Access to Opportunity and Safety for Vulnerable Pedestrians
Tabitha Combs, University of NC Department of City & Regional Planning
An AV dominant transportation system may ultimately yield net positive benefits with respect to safety and mobility. However, there will be downsides accompanying this shift in how our transportation system functions. Absent appropriate interventions from planning & policy, these downsides will accrue disproportionately to lower-resourced populations—including but not limited to a failure to substantially reduce pedestrian deaths, and further worsening of transportation inequities already in place today.

Based on findings from a recent paper published in the American Journal of Preventive Medicine [8], Dr. Tabitha Combs built the case that pedestrian safety benefits of AVs are unlikely to reach the most vulnerable and least resourced members of the population for at least several decades. This is due not necessarily to technical limitations in pedestrian detection and avoidance, but to spatial mismatches in where and how AVs will operate, and where and to whom pedestrian fatalities occur. In the meantime, stop-gap solutions to improve pedestrian safety and enhance the efficiency and performance benefits of AVs will increase barriers to access for those same populations—children, the elderly, those with cognitive challenges, and those with limited resources.

Limiting access to opportunity leads directly to greater risk-taking by vulnerable populations, either with respect to pedestrian behavior (e.g., hopping over physical barriers to jaywalk across busy streets) or foregoing travel and participation in critical out-of-home activities. Programs exist to help bridge access gaps in the transportation system now, such as subsidized transit and regulation of taxi fares, but AVs will soon introduce dramatic changes in these markets. Things like access to curb space for pickups and drop-offs, the ability to link into multi-car platoons, and even the code governing how cars operate may all contribute to new ways people will have to compete for accessibility. Policy-makers and regulatory agencies thus far have been reluctant to anticipate and address accessibility discrepancies that will inevitably emerge from these changing markets. This reluctance means more vulnerable people will be shut out of the AV mobility market at the same time as their access to opportunity on foot erodes, forcing them into ever riskier tradeoffs.

These tradeoffs can be averted, but only with appropriate, proactive policy intervention to establish and regulate standards for AV operation that prioritize the needs of pedestrians, and ensure uninhibited access to opportunity by foot becomes the central focus of transportation safety.

3 Panel Discussion: The Role of Equity Considerations in AV/VRU Interactions

In this panel, presenters each introduced their work and perspectives on how the proliferation of AVs may impact equity and safety, particularly ways in which city planning can both take advantage of AV technology and potential hurdles and barriers to equity. Following the brief presentations, panelists engaged in a roundtable discussion with each other and with the audience. In this section we summarize these initial remarks and the following discussion with the audience.

The panel was kicked off by Bernard Soriano, Deputy Director of the California Department of Motor Vehicles (CA DMV). Dr. Soriano briefly described how the CA DMV is preparing for AV technology deployment and is working with companies to bring AVs safely onto public roads. California has had regulations in place on AV technology since 2014, with 56 companies currently approved for testing. Six-hundred vehicles are currently being tested on California roadways, with approximately 1600 test drivers. As of the date of this panel there had been 79 crashes of vehicles with AV technology in California, many of which involve rear-end collisions by vehicles under the control of human drivers. He also discussed that California has just begun the process of issuing permits for completely driverless testing, with two companies currently in the review process to do so.

Following Dr. Soriano's presentation, Dr. Sarah McCullough from the Feminist Research Institute at UC Davis presented her perspective on VRUs, mobility justice and transportation equity. This perspective is informed by her research on mobility justice and participation in a multiracial collective of transportation professionals called The Untokening [9]. The Untokening shares professional practices and resources that center the lived experiences of marginalized communities to address mobility justice and equity. The Principles of Mobility Justice [10] that this collective offers provides an important resources for those interested in ensuring that the emergence of autonomous vehicles addresses equity issues.

There is a history of deep inequity built into our transportation infrastructure and cycles of innovation. More advanced technologies such as light rail and underground systems tend to disproportionately benefit suburban communities rather than urban neighborhoods [11]. White flight from urban communities fueled the growth of these suburban neighborhoods, even as Blacks and people of color were restricted to purchasing homes in redlined neighborhoods [12]. This practice of redlining resulted in racial segregation and environmental injustices, whose legacies still live with us today [13]. A meaningful engagement with equity requires us to acknowledge these histories and consider how new transportation investments and innovations, like AVs, can most benefit those historically neglected.

An equity approach to AVs and vulnerable road users also suggests that we broaden our definition of safety. Often, discussions of safety in transportation are limited to collisions. In practice, street safety also encompasses issues such as over-policing, harassment, theft, detainment, and accessibility. The Principles of Mobility Justice suggest that safety should be defined by the most socially, economically, and legally vulnerable. If AVs are to have a meaningful impact on transportation inequalities, then they must consider how emerging technologies impact these diverse, yet interlocking issues of safety. Even better, the development of new technological innovations will be primarily driven by these concerns.

Equity is a sociocultural problem, and AVs are a sociocultural technology. We need sociocultural innovation as much as or more than technological innovation if we hope to move the needle in a meaningful way toward reducing transportation inequalities. We must seriously consider the following questions: What problems are AVs addressing? Are these the problems of marginalized communities? What are the primary difficulties that these communities face? How might these issues be addressed or exacerbated by the emergence of AVs?

As many have pointed out, we are on the brink of another transportation revolution. Historically, such revolutions have both benefited and disadvantaged those at the margins of society. The bicycle contributed to white women's emancipation, while the proliferation of highways fostered mass racial segregation and urban disinvestment. For AVs to have a deep social benefit requires sociocultural innovation and deep engagement with marginalized communities. It requires working with environmental justice advocates, and movements responding to other community safety issues such as police shootings, ICE raids, family separation, and #MeToo conversations. We must ask, who will be affected by this, and is not in the room?

Ryan Snyder, a transportation consultant with the Transpo Group, then presented his perspective on AV policy and equity. Mr. Snyder's comments drew primarily from his recent paper *Public Health and Equity Considerations of Autonomous Vehicles in California* [14], which form the basis for this summary. The California Department of Public Health defines "health equity" as "efforts to ensure that all people have full and equal access to opportunities that enable them to lead healthy lives." The paper explored various future scenarios to assess how AVs will impact health equity in California. AVs will be extremely disruptive technology. They have potential to create more equitable communities with improved access, to reduce greenhouse gases and to make our communities healthier and more livable. At the same time, AVs may exacerbate inequities, undermine conditions for walking and bicycling, increase greenhouse gas emissions, and render our communities less livable. The difference will be the public policies that are enacted.

Since full-operation AVs don't yet exist on public roadways, we are left to speculate. However, this paper documents research that can be applied to take educated guesses at what outcomes may arise as a result of various public policies. It defines the related health outcomes that are most likely to be impacted by AVs as:

- Impact on active transportation
- Greenhouse gas emissions and air pollution
- Traffic safety
- Mental health
- Possible Impacts of 5G wireless technology

The paper identifies possible health equity considerations as:

- Accessibility
- Job losses from automation

- Exposure to traffic and associated impacts
- In-vehicle personal safety

This paper declares that the primary determinants of the health and health equity impacts of AVs will be:

1. The degree to which transportation is shared
2. The degree to which transportation is electrified, and the sources of electrical generation
3. Other miscellaneous public policies

The paper assesses three policy scenarios and what their impacts will most likely be. The first scenario evaluates a hands-off approach. The second looks at what would result from modest policies. The third weighs the outcomes of assertive policy. The paper concludes that the most assertive policy scenario will result in the most optimal outcomes for health and health equity. It closes with policy recommendations to reach the most favorable outcomes.

The panel presentations were concluded by Robert Wall Emerson, a professor in the Department of Blindness and Low Vision Studies at Western Michigan University. His comments drew from his expertise in the area of low vision, and consider the opportunities and impact of automation on people with visual impairments.

For much of history, the long cane (or something like it) has been the most consistent, reliable mobility tool for people who are blind. Since the 1940s there have been dozens of electronic mobility aids designed to assist people who are blind in their navigation and/or mobility but few have been very successful or widely used. The advent of GPS technology changed this dynamic, giving people who are blind access to real time location information with which they could make travel planning and decisions. Dedicated GPS based devices have quickly been supplanted by devices such as smart phones that incorporate GPS based functionality but also offer a range of other mobility related information or applications. Even so, the physical long cane remains a needed tool to support mobility for people who are blind by providing information about the immediate physical environment.

What has been missing in this development of technology to assist people who are blind in their mobility is truly accessible transport for independent mobility over long distances. If a person who is blind is traveling somewhere that is too far to walk, they must engage the service of a driver or access some sort of transportation system (Uber, Lyft, bus, metro, train, etc.). This can lead to inconveniences with scheduling, long wait times, unreliability in service, and inconsistency in performance. Connected and autonomous vehicle technology holds the promise of offering a person who is blind the chance to engage with a transportation system that is available on their command and is under their control for the duration of the travel.

However, there are certain obstacles that must be overcome before the idealized version of access is realized. Autonomous and connected vehicles need to be designed in order for people who are blind to be able to use them effectively. To do this, a vehicle must be able to announce to a waiting passenger that it has arrived and where it is in a manner that allows the blind traveler to easily find the vehicle and enter. The vehicle needs to be able to accept input from a person who is blind and communicate during travel what is happening in the environment, where the vehicle is, time to the destination, and any other relevant information the passenger may want or need to know. Finally, the vehicle must be able to communicate to the passenger not only that they have arrived but what the immediate environment is like so that the passenger need not spend an inordinate amount of time tactually exploring the environment before figuring out where they need to walk after disembarking.

The promise that connected and autonomous vehicle technology holds for all members of society who currently do not drive is huge. But there is the very real possibility that social inequities that currently exist in society will only be exacerbated by the deployment of such technologies. If a segment of the population lives in an area that is not well serviced by public transportation, will they be well serviced by newer and more sophisticated transportation? If a segment of the population has limited access to transportation and technology in general due to financial constraints, it is quite possible that deployment of sophisticated and expensive autonomous technologies will not be a daily part of their world. People with disabilities, including those who are blind, are disproportionately represented in segments of the population that are more poor and have limited access to services. The way that new travel technologies are deployed will impact how much of the population they will actually serve.

4 Discussion and Action Items/Research Needs

Following the brief perspective presentations of each panelist, the floor was opened to questions and discussion between the moderator, panelists, and the audience. This section will present a selection of the questions posed and synopses of the discussion generated.

What equity issues surround the loss of jobs?
AVs have the potential to contribute to both job gains (for example, for people with disabilities who might not otherwise be able to drive) and job losses (among people who work in industries that may be bypassed by automation and/or people who cannot afford automation technology). This question generated quite a bit of discussion surrounding regulations that exist and could potentially be drafted; in what cases could policy redistribute risk, and when is it appropriate to do so? Automation may also accelerate job loss in specific at-risk communities, including those in rural areas that are already suffering economic harm from the shift from a manufacturing toward a service-based economy. New technologies may also pose risks to people who are less able to adapt to new ways of doing things, including older workers and those with less education. One potential method of distributing risk is to shift incentives; for example, by replacing the gas tax with a mileage tax or a fee on empty seats. There was also considerable discussion in this question of the role of federal and state governments, and which is responsible for what aspects of automation development.

In what ways might AVs affect racial equity and policy development? How do we make room for new voices?
An important topic of discussion was how to include the voices of people, particularly minorities and people with low SES, who may not be adequately represented in the

government or industry sectors responsible for AV development and implementation. It is important for members of the planning and development community to connect with grassroots community movements with expertise in areas related to and overlapping with mobility. This engagement could involve substantial resources, which should be considered at the highest levels to ensure community input and buy-in as technology progresses. Further, it could be beneficial to take a public health approach, including outreach designed to both inform and learn. One important point is that OEMs must be convinced that all communities must be involved for automation technology to be accepted, which could be helped by increased industry representation at conferences and meetings such as AVS.

Given most of this technology is currently on more expensive vehicles, how do we (and should we) prioritize these conversations for communities with much more pressing needs?
This question is difficult, because many communities have much more immediate needs to focus on than technology that is several years or decades away and may not appear to provide much near-term benefit. Similar to the previous question, conversation returned to the idea that the transportation and planning communities need to engage with local community groups and find ways to synergize interests. For example, with a shift to automation may come a shift in the traditional vehicle ownership model to one of shared service; this could provide new transportation opportunities for communities in general and in particular for communities that may not currently be having their transportation needs met. However, this notion was met with some skepticism, as ride-share services still struggle to meet the needs of low-SES communities and tech companies such as Google and Facebook are meeting resistance from local communities when they introduce employee-only bussing services [15].

Finally, the group resolved a set of action items and research needs that should be addressed to support the continued focus on equity as AV systems are deployed on our roadways. These included:

- Research needed to evaluate on-road as well as trip-level behavior
- Recognize the tradeoff & interdependence between safety & access to transportation
- Explore ways to shift the culture to prioritize pedestrians, maximize access and safety
- Identify needs & risks for people with disabilities in particular, within & outside vehicles
- Find ways to engage at-risk communities including low-SES areas
- Recognize & learn from the history of how transportation has affected communities, particularly low SES
- Work with stakeholders at all levels from community to OEMs, bring in government at all levels – local, state & federal - to identify & solve problems before they become serious and/or permanent.

In conclusion, this session provided an opportunity for experts in transportation and automation technology to open discussion with experts in equity, disability studies, planning, and mobility for vulnerable road users. Important topics were introduced that must be pursued for automation to succeed across a broad range of users and locales.

References

1. Brown, J.: A tale of two visions: Harland Bartholomew, Robert Moses, and the development of the American freeway. J. Plann. Hist. **4**(1), 3–32 (2005). https://doi.org/10.1177/1538513204272856
2. Semuels, A.: The Role of Highways in American Poverty (2016). https://www.theatlantic.com/business/archive/2016/03/role-of-highways-in-american-poverty/474282/
3. Semuels, A.: A Departure from Decades of Highway Policy (2016). https://www.theatlantic.com/business/archive/2016/03/the-transportation-secretary-speaks-out-against-highways/475749/
4. Owens, J.M., Sandt, L., Morgan, J.F., Sundararajan, S., Clamann, M., Manocha, D., Steinfeld, A., Maheshawri, T., Cooper, J.: Challenges and opportunities for the intersection of vulnerable road users (VRU) and AVs. In: Road Vehicle Automation 5. Springer, Cham (2018)
5. Cook, S., Englund, C., Holmgren, M.: Automation for increased accessibility? Report, Strategic Innovation Program Drive Sweden, Gothenburg, Sweden (2017). https://www.drivesweden.net/sites/default/files/content/final_report_-_automation_for_increased_accessibility_-_excluding_appendices.pdf. Accessed 24 Aug 2018
6. United Nations: Convention on the Rights of Persons with Disabilities (2006). http://www.un.org/disabilities/documents/convention/convention_accessible_pdf.pdf. Accessed 24 Aug 2018
7. Feng, F., Bao, S., Hampshire, R.C., Delp, M.: Drivers overtaking bicyclists—an examination using naturalistic driving data. Accid. Anal. Prev. **115**, 98–109 (2018)
8. Combs, T.S., Sandt, L.S., Clamann, M.P., McDonald, N.C.: Automated vehicles and pedestrian safety: exploring the promise and limits of pedestrian detection. Am. J. Prev. Med. **56**(1), 1–7 (2019)
9. http://www.untokening.org
10. http://www.untokening.org/updates/2017/11/11/untokening-10-principles-of-mobility-justice
11. Wells, C.W.: Car Country: An Environmental History. University of Washington Press, Seattle (2012)
12. Hernandez, J.: Redlining revisited: mortgage lending patterns in Sacramento 1930–2004. Int. J. Urban Reg. Res. **33**, 291–313 (2009)
13. Pulido, L.: Rethinking environmental racism: white privilege and urban development in Southern California. Ann. Assoc. Am. Geogr. **90**, 12–40 (2004)
14. Snyder, R.: Public Health and Health Equity Considerations of Autonomous Vehicles in California. Transpo Group, July 2018. http://www.transpogroup.com/publichealthandavs.html
15. https://www.nytimes.com/2018/05/31/us/google-bus-protest.html

Part III Vehicle Systems and Technology Development

Evaluation of Automated Driving by Large-Scale Piloting on European Roads – The L3Pilot Project

Christian Rösener[1(✉)], Adrian Zlocki[2], Hendrik Weber[1], and Johannes Hiller[1]

[1] Institut für Kraftfahrzeuge (ika), RWTH Aachen University, Steinbachstraße 7, Aachen, Germany
{christian.roesener,hendrik.weber,johannes.hiller}@ika.rwth-aachen.de
[2] fka Forschungsgesellschaft Kraftfahrwesen mbH Aachen, Steinbachstraße 7, Aachen, Germany
adrian.zlocki@fka.de

Abstract. Automated driving is still the main focus of research activities in the European automotive industry. The European Commission's flagship project "L3Pilot" is testing the viability of automated driving as a safe and efficient means of transportation and additionally exploring and promoting new service concepts to provide inclusive mobility. Therefore, L3Pilot will create a standardized Europe-wide piloting environment, coordinating activities across the piloting community in order to acquire data from 100 vehicles with 1000 drivers operating in 10 countries in Europe. Concluding, the project will evaluate the automated driving functions (ADFs), calculate the impacts and provide a cost-benefit analysis.

The evaluation of all ADFs is challenging and requires a sophisticated methodology. This paper describes the overall evaluation approach and in particular the methods for technical- and impact assessment that are taken in the project.

Keywords: L3Pilot · Pilot · Evaluation methodology · Technical assessment · Safety impact assessment

1 Introduction

Automated driving technology has matured to a level motivating an extensive phase of road tests, which can answer the key questions on safety, security, interaction, and the societal benefit before market introduction. A large-scale pilot provides appropriate assessment of the impacts of automated driving. "What is happening inside and outside the vehicle?" and "How can vehicle security be ensured?" are two questions the L3Pilot is focused on, as well as the evaluation of the societal impact and emerging business models. L3Pilot is a European research project funded by the European Commission with €36 million of funding. The project started in September 2017 and has a duration of 48 months.

G. Meyer and S. Beiker (Eds.): AUVSI 2019, LNMOB, pp. 75–83, 2019.
https://doi.org/10.1007/978-3-030-22933-7_8

The core of the project is a large-scale pilot in which 100 automated vehicles will be operated on European roads collecting subjective and objective data, which is used to derive predictions on the social, economic and ecological impact of automated driving. Many partners of the consortium can rely valuable experience from predecessor projects like AdaptIVe [1], DRIVE C2X [1] and euroFOT [3], which form an optimal basis for the implementation of the experiment from a technological point of view as well as from a methodological one.

The overall objective of the L3Pilot project is to test and study the viability of automated driving as a safe and efficient means of transportation and to explore and promote new service concepts to provide inclusive mobility. In order to achieve this objective, a standardised Europe-wide piloting community will be created within which the piloting activities will be coordinated and harmonised. By this means, it will be possible to pilot, test and evaluate ADFs and connected automation. Furthermore, efforts will be made to innovate and promote ADFs for market introduction and wider awareness.

Overall, the project is expected to have impact on various areas. Both technical and methodological knowledge will be generated, which allow to derive requirements for function design and support simulative testing. Furthermore, an understanding of the societal impacts of automated driving will be achieved. These concern possible gains in road safety, reductions of emissions and influences on infrastructure, jobs, the economy and healthcare. The business impact will consist of guidelines defining a common basis for system design as well as validation. User data collected from the pilot will serve to explore possible business cases for market introduction of automated driving. A deployment roadmap will give an overview of necessary actions that need to be undertaken by various parties involved in automated driving in order to deploy automated driving on European roads smoothly.

2 Evaluation Methodology

Tests on public roads in real traffic are essential for the evaluation of SAE L3 and SAE L4 [4] systems since all relevant aspects of an ADF are addressed. In addition, tests on public roads offer a high complexity and variety of driving situations. Large-scale testing efforts in public traffic can ensure that the situations in which a system is tested represent all relevant driving situations. The required distance to be driven in order to guarantee a safe performance of an ADF has been predicted as high as ten million kilometres [5]. For tests conducted within the design and development stage at the manufacturer the collected data is kept confidential, especially data on critical situations involving automated vehicles. Due to this lack of data, it is not possible to draw conclusions of the overall impact that the introduction of the ADF will have. Even if data were shared, an efficient evaluation of data from multiple players would be problematic since data formats are not harmonised.

Within L3Pilot, data for evaluation will be collected from different European manufacturers, vehicles and automation functions. Although not all data is shared on all detail levels the evaluation possibilities of the available data will exceed all known data sources up to today. This will allow L3Pilot to provide new insights for the market introduction of L3 automated driving.

The methodology applied in L3Pilot follows the guidelines defined in the FESTA handbook [6]. This handbook gives a well-elaborated roadmap for measures to be taken in preparation of a field operational test and during its deployment as well as under what aspects the gathered data should be evaluated. Based on these guidelines four key phases of the L3Pilot project can be identified as shown in Fig. 1. A distinction is made between the stages "PREPARE" (i), "DRIVE" (ii), "EVALUATE" (iii) and legal aspects & cyber-security (iv).

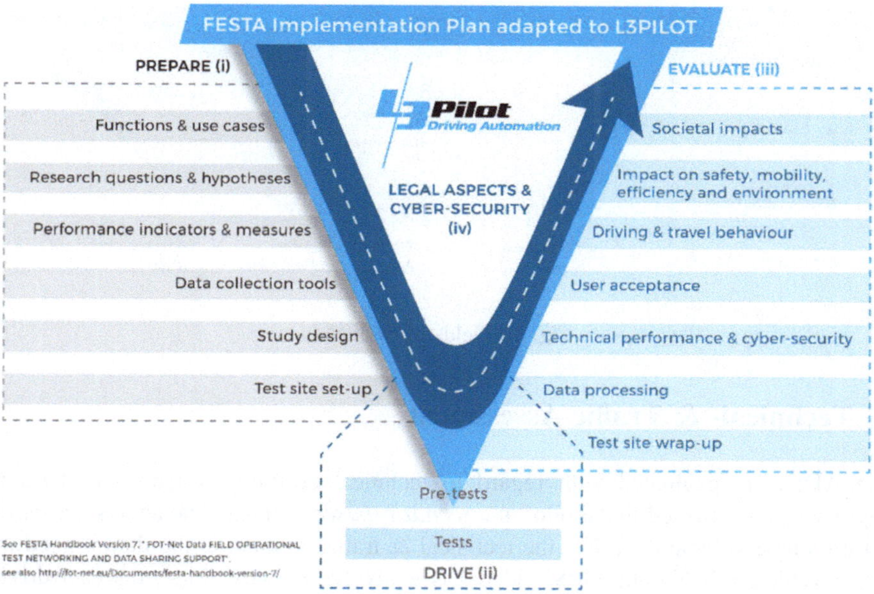

Fig. 1. Structure of L3Pilot project in accordance with FESTA guidelines.

During the "PREPARE" stage, research questions and the respective hypotheses are defined to assess the use cases of the project. In L3Pilot, the use cases are traffic jam, motorway (including traffic jam), parking and urban automation. Data collection tools are developed that are capable to analyse the derived performance indicators (PIs). In order to make sure that data to answer all research questions is collected, a harmonized study design is developed for all pilot sites. Afterwards, the subjective and objective data is collected during the "DRIVE" phase. This data is assessed in the "EVALUATE" phase. Similar to previous projects like PReVAL [7] and AdaptIVe [8], the data is assessed in four different areas. These are technical & traffic-, user-, impact- and socio-economic evaluation. As illustrated in Fig. 2, the entire evaluation is based on real-world data collected during the pilot. While technical & traffic evaluation and user evaluation are in-depth analyses of the pilot data, the impact evaluation is based on aggregated and thus de-identified results of this analysis. In consequence, the impact in terms of traffic safety and efficiency will be derived based on real-world driving data. In the following, the evaluation approach is presented following the example of technical & traffic- as well as impact evaluation.

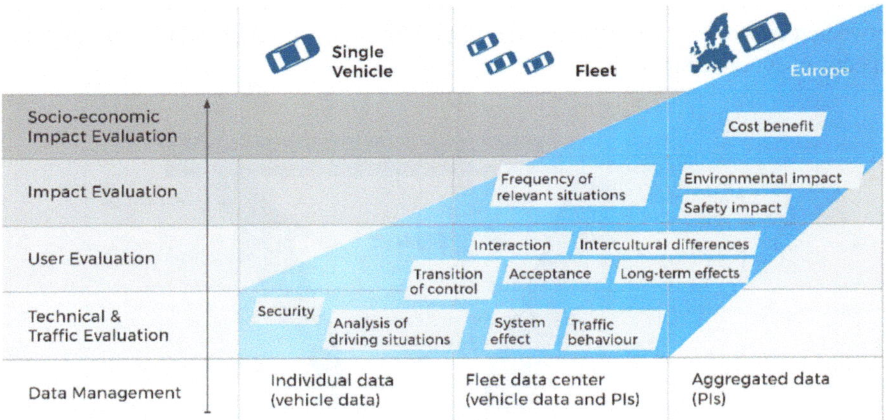

Fig. 2. Fields of evaluation

3 Technical- & Traffic Assessment

The ADFs are evaluated with regard to technical & traffic aspects based on the objective data collected in the pilot in a *scenario-based* manner. The analysis is carried out on single vehicle data. For the technical & traffic evaluation the data logged in a single vehicle (CAN-data, GPS, videos) is analysed stepwise. First, relevant driving scenarios are automatically detected. The performance indicators (PIs) are calculated for each identified driving scenario. In the last step, the derived PIs are interpreted in order to answer the defined research questions and hypotheses. Technical & traffic evaluation cover the following areas:

- What is the system's technical performance?
- What is the impact on the own driving behaviour?
- What is the impact of ADF on the interaction with other road users?
- What is the impact on the behaviour of other traffic participants?

In this section, the evaluation methods for technical & traffic are introduced. A distinction is made between four different groups of ADFs: motorway including traffic jam, traffic jam, parking and urban. The purpose of this distinction is to group and aggregate the results of different ADFs.

Since the operational design domain of ADFs covers a high dimensional situation space including many different driving situations with lots of variations, the assessment approach must ensure that data of the reference and the test object is available in a sufficient amount. Therefore, a holistic assessment approach covering as many different driving situations as possible is needed. The authors propose a scenario-based assessment approach based on real driving data of the test and the reference driving behaviour in accordance with [9]. By using this approach, the different test scenarios and variations of the test scenarios are generated stochastically by real-world traffic dynamics. As depicted in Fig. 3, the developed methodology first foresees a classification of test and reference driving behaviour data in relevant scenarios.

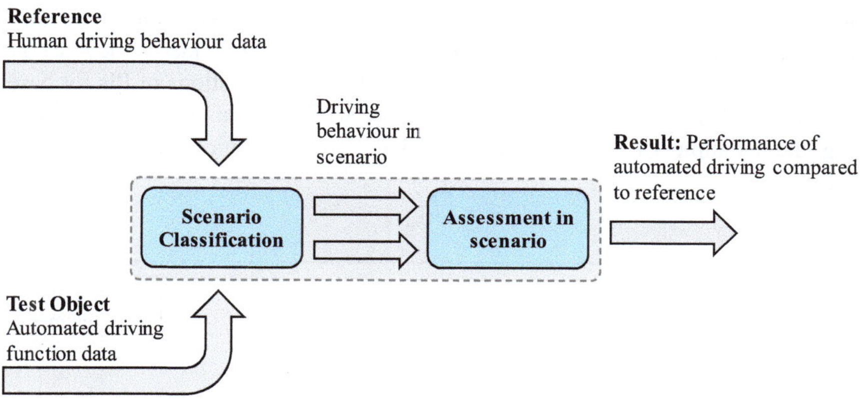

Fig. 3. Schematic view of method for technical & traffic assessment

Due to the diverse characteristics of traffic, the test approach must ensure that sufficient test and reference data is available. For this purpose, parts of the euroFOT database are considered [3] for estimating the mean frequencies of relevant driving scenarios. For calculating the minimal test distance for the occurrence of k = 30 driving scenarios which are necessary to assess the function, a cumulative Poisson distribution is assumed. Based on the mean distance necessary for the occurrence of a single event s_{ref}, the necessary distance is calculated for the occurrence of k events with a probability of P = 95%. The basis for the calculation of the minimum distance is given with the following equation describing the Poisson distribution, where the probability for the occurrence of a driving situation is given by:

$$P = \sum \frac{\lambda^k}{k!} e^{-\lambda}$$

The expectancy value can be obtained by:

$$\lambda = \frac{s_k}{s_{ref}}$$

The resulting test distances are listed in Table 1.

Table 1. Estimated test distances.

Driving scenario	Test distance			
	k = 5	k = 10	k = 20	k = 30
Cut-in	250 km	350 km	600 km	800 km

After collecting the data in a reasonable amount, for both baseline and treatment data an enrichment is applied. In this step, all derived measures (DMs) related to dynamic objects in the environment of the ego vehicle are computed. Afterwards incidents and driving scenarios are detected based on defined thresholds. The detected incidents and driving scenarios are validated in the next step by video review according to the coding scheme presented in [10]. As a result, the distributions of PIs for baseline and treatment data are computed for each driving scenario. This is illustrated in Fig. 4 taking the example of the PI "time headway" for a "vehicle following" driving scenario.

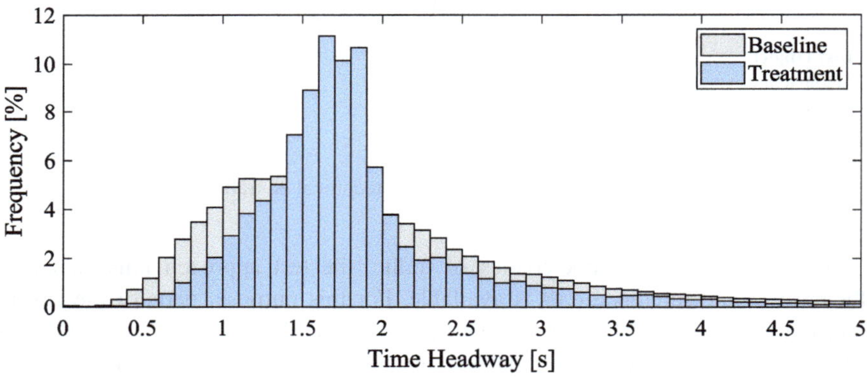

Fig. 4. Exemplary illustration of histogram of performance indicator "time headway" for baseline and treatment.

After classification of the relevant driving scenarios and derivation of the respective PIs, the earlier defined hypotheses are evaluated. For determining whether the behaviour of the ADF is within the range of normal driving behaviour, and furthermore to quantify the deviation from normal driving behaviour, an appropriate method has to be identified. In this case hypothesis testing cannot be used due to the large test samples obtained. Therefore, the usage of the quantitative measure 'effect size' is proposed in this approach, which is, according to [11], a simple way of quantifying the difference between two groups, that reveals many advantages over the use of tests of statistical significance alone. As depicted in [11], the effect size is a standardized mean difference between two groups and emphasizes the size of the difference rather than confounding this with sample size. The effect size d is calculated in order to estimate the deviation of the behaviour of the ADF compared to human driving behaviour, see equation below:

$$d = \frac{\mu_{experimental} - \mu_{reference}}{\sqrt{\frac{\sigma^2_{experimental} - \sigma^2_{reference}}{2}}}$$

4 Safety Impact Assessment

The safety impact assessment investigates the changes in accidents and injuries in road traffic due to automated driving. While in the past active safety systems were assessed based on a set of recorded accident scenarios obtained from human driving [12], this approach will not be sufficient concerning ADFs. ADFs – in contrast to active safety systems – continuously control the behavior of the vehicle. Due to this reason, it is possible that ADFs do not get involved in previously important accident scenarios any longer while other, for human driving less relevant accident scenarios, become more important.

However, it can be assumed that the relevant driving scenarios leading to certain accident scenarios will not change with automated driving. Their frequency and severity may rather change with automated driving [13]. For this reason, besides re-simulation of detailed accident scenarios for identifying the changes in severity due to automated driving, the changes in frequency of occurrence of relevant driving scenarios are investigated based on traffic simulations and the results of the technical & traffic evaluation from L3Pilot. The steps of the safety impact assessment are elaborated in the following:

- **Description of ADF & Identification of the effectiveness field**
  Based on the operational design domain of the ADFs the target population of addressed accidents is identified in the accident statistics. For example, a Motorway-Chauffeur may address about 53% of all accidents on German motorways [13].
- **Changes in frequencies of driving scenarios**
  Since ADFs operate continuously their engagement may lead to a change in the frequency of occurrence of certain driving scenarios, e.g. cut-in. This is investigated based on traffic simulations and the results of the technical & traffic evaluation of the pilot data.
- **Changes in severity of driving scenarios**
  Within the driving scenarios which are relevant for the ADF its performance is compared with human driver reference performance.
- **Scaling-up of effectiveness to national target level**
  Finally, the identified effectiveness fields in the accident statistics are used to scale-up the previously identified effects.

The overall approach for safety impact assessment incorporating the prediction of frequencies of driving scenarios is presented in Fig. 5.

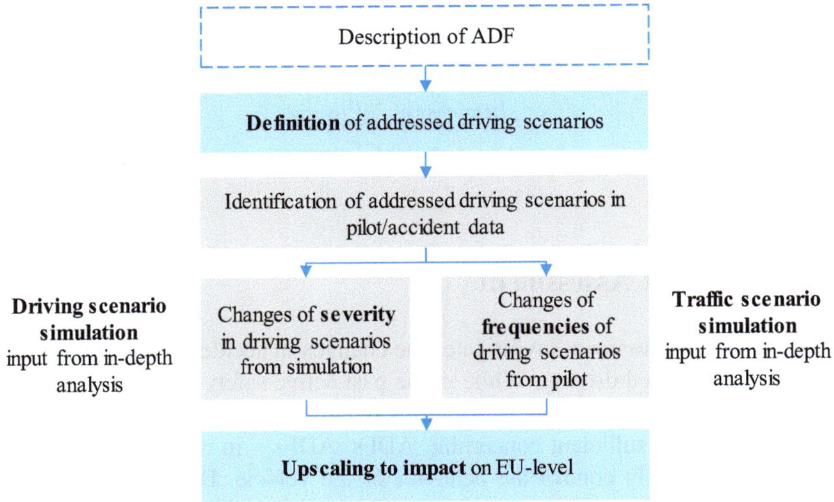

Fig. 5. Schematic of method for safety impact assessment based on [13].

5 Summary and Outlook

The overall objective of the L3Pilot project is to test and study the viability of automated driving as a safe and efficient means of transportation by a standardised Europe-wide piloting community will be created. Within this community the piloting activities will be coordinated and harmonised.

This paper describes the evaluation approach that is applied in the European research project L3Pilot following the example of technical & traffic assessment and safety impact assessment. A major challenge is the variety of ADFs ranging from motorway- to urban automation functions. In order to cope with this variety, a scenario-based assessment approach is established that is generating the results for each driving scenario instead of for each ADF. Afterwards, these can be aggregated for the analyzed ADFs.

Next, the assessment of the impact in terms of traffic safety poses challenges due to the large situation space addressed by complex ADFs. Although 100 automated vehicles will be assessed within the L3Pilot project, the penetration rate of automated vehicles will be too low to draw conclusions on overall traffic safety. To cover the entire situation space and to be able to draw conclusions on high penetration rates of automated vehicles a safety impact assessment will be performed. To ensure its validity, it is based on the aggregated results of the in-depth analysis within L3Pilot and is complemented by traffic simulations of automated vehicles.

Acknowledgment. The research leading to these results has received funding from the European Commission Horizon 2020 research programme under the project L3Pilot, grant agreement number 723051. Responsibility for the information and views set out in this publication lies entirely with the authors. The authors would like to thank all partners within L3Pilot for their cooperation and valuable contribution.

References

1. Knapp, A., Fahrenkrog, F., Zlocki, A.: Automated Driving Applications and Technologies for Intelligent Vehicles – AdaptIVe, 23rd Aachen Colloquium Automobile and Engine Technology (2014)
2. Stahlmann, R., Festag, A., Tomatis, A., Radusch, I., Fischer, F.: Starting European field tests for Car-2-X communication: the DRIVE C2X framework. In: 18th ITS World Congress and Exhibition (2001)
3. Benmimoun, M., Fahrenkrog, F., Zlocki, A., Eckstein, L.: Incident detection based on vehicle CAN-data within the large scale field operational test "euroFOT". In: 22nd Enhanced Safety of Vehicles Conference (ESV 2011). Washington, DC, USA (2011)
4. N.N. Taxonomy and Definitions for Terms Related to On-Road Motor Vehicle Automated Driving Systems, SAE Standard J3016_101401 SAE International (2018)
5. Winner, H.: Quo vadis, FAS? In: Handbuch Fahrerassistenzsysteme. Springer Fachmedien Wiesbaden (2015)
6. FESTA Consortium: "FESTA Handbook" FESTA Support Action, Field opErational teSt supporT Action. D6.4 (2008)
7. Scholliers, J., Hendirks, F., Aust, M.L., et al.: Project final report and recommen-dation for future assessments. PReVAL project, Deliverable DL IP D12/D16.4 (2008)
8. Fahrenkrog, F., Rösener, C., Zlocki, A., et al.: Technical evaluation and impact assessment of automated driving. In: Road Vehicle Automation 3. Springer Lecture Notes in Mobility (2016)
9. Rösener, C., Fahrenkrog, F., Uhlig, A., et al.: A scenario-based assessment approach for automated driving by using time series classification of human-driving behaviour. In: IEEE Intelligent Transportation Systems Conference (ITSC), Rio de Janeiro (2016)
10. N.N. Researcher's Dictionary for Video Reduction Data, SHRP 2, Virginia Tech Transportation Institute, Virgina, USA (2015)
11. Coe, R.: It's the effect size, stupid: what effect size is and why it is important. In: Annual Conference of the British Educational Research Association, pp. 12–14. University of Exeter, UK (2002)
12. Noort, M.v., Bakri, T., Fahrenkrog, F.: SIMPATO - the safety impact assessment tool of interactive. In: 16th International IEEE Annual Conference on Intelligent Transportation Systems. The Hague (2013)
13. Rösener, C., Hennecke, F., Sauerbier, J., et al.: A Traffic-based Method for Safety Impact Assessment of Road Vehicle Automation, 12. In: Uni-DAS Workshop Fahrerassistenzsysteme und automatisiertes Fahren, Walting, Germany (2018)

New Advances in Intelligent Intersection Management with Connected and Automated Vehicle Technology

Mehdi Zamanipour[1(✉)], Yiheng Feng[2], and Govind Vadakpat[3]

[1] National Research Council, Federal Highway Administration, 6300 Georgetown Pike, McLean 22101, USA
mehdi.zamanipour.ctr@dot.gov

[2] Transportation Research Institute, University of Michigan, 2901 Baxter Rd., Ann Arbor, MI 48109-2150, USA
yhfeng@umich.edu

[3] Federal Highway Administration, 6300 Georgetown Pike, McLean, VA 22101, USA
g.vadakpat@dot.gov

Abstract. Considerable research studies coupled with several deployment projects have been conducted recently to investigate potential effects of different cooperative automation technologies in controlling signalized junctions. The focus has been on how vehicles and infrastructure can cooperate toward safer and more efficient intersection operations. In this chapter, a brief review of some ongoing research projects as well as real world implementations that were presented during the Automated Vehicles Symposium (AVS) 2018 are discussed. The review includes the specifications of the projects and the challenges in implementation of the new technology. Three of the near future possible deployments are presented as well.

Keywords: Connected and automated vehicle (CAV) · Intersection control management · Traffic signal control

1 Introduction

Traditional approaches to intersection management are less intelligent and the control decisions are made without awareness of the entire state of the traffic. Typically, intersections are operated in a coordinated manner with a number of signal timing plans that have been designed and allocated to accommodate different volumes, based on time-of-day and day-of-week traffic flow patterns. These timing plans are not updated very often due to the high cost of data collection and analysis. Although vehicle actuated signal controls are more efficient than fixed-timing signal operations, they are not inherently designed to optimize traffic users delay and/or fuel consumption in real time. Even in more advanced traffic management systems such as adaptive control, the lack of reliable traffic user detection systems causes inefficiencies. With advances in connected vehicle technologies and dedicated short range communication (DSRC), different traffic modes can communicate to each other. Through vehicle-to-X (V2X)

G. Meyer and S. Beiker (Eds.): AUVSI 2019, LNMOB, pp. 84–91, 2019.
https://doi.org/10.1007/978-3-030-22933-7_9

communications, vehicles can exchange information with each other and with infrastructure. This environment provides higher traffic resolution data and will result in a well-informed decision making process. In this smarter inter-connected environment, safety, mobility, and environmental performance measures can be improved.

Recently, several research projects have been implemented in different U.S. cities to investigate the effects of connected and automated vehicle (CAV) technology in intersection control and management. Most of these projects were presented during the AVS 2018 conference, breakout session No. 13, "New innovations in intersection control with cooperative automation". In the following sections, a brief review of completed, ongoing and potential future projects are presented. The outcomes, lessons learned as well as top challenges are discussed.

2 Completed Projects

2.1 Multi Modal Intelligent Traffic Signal System (MMITSS)

The Multi-Modal Intelligent Traffic Signal System (MMITSS) project is part of the Connected Vehicle Pooled Fund Study (CV PFS) entitled "Program to Support the Development and Deployment of Connected Vehicle System Applications." The CV PFS was developed by a group of state and local transportation agencies and the Federal Highway Administration (FHWA). MMITSS consist of four major components: Intelligent Traffic Signal System (I-SIG), Transit and Truck Signal Priority (TSP), Real-time Performance Monitor (RTPM), and Mobile Accessible Pedestrian Signal System (PED-SIG).

This project is conducted by the University of Arizona (PI) and the PATH program at the University of California, Berkeley in two different testbeds. The requirements of the MMITSS system to be implemented in the California testbed are different than those for the Arizona Test Bed. The primary differences are in the hardware architecture, including traffic signal controllers used in each testbed, algorithms for signal priority and intelligent signal control. The California testbed utilizes the Caltrans Type 2070 controllers with AB3418 protocol over serial RS-232 communications. The Arizona testbed utilizes Econolite ASC/3 and Cobalt controllers with NTCIP (National Transportation Communications for ITS Protocol) over Ethernet communications. Each of these controllers provides different signal timing logic (control software) and requires different communications interfaces [1].

In terms of software design, although there are many common components, but the core software architecture and control algorithms are different. The Arizona testbed implemented adaptive signal control and the California testbed chose to enhance the existing coordinated-actuated signal control with a common cycle length. As a result, the algorithms implemented for intelligent signal control and signal priority are different for the two testbeds [2–4].

The field data analysis demonstrated that MMITSS applications effectively improved the travel time and reduced delay for the DSRC equipped vehicles. In particular, the signal priority component reduced delay of connected trucks by up to 20% and the I-SIG improved travel time reliability by up to 56%, compared to the base case.

The simulation study found that I-SIG achieved vehicle delay reductions up to 35% and TSP effectively saved travel time for both transit and passenger vehicles on the corridor where TSP was operated; but occasionally increased the system-wide delay, due to reduced green times on the side streets [5]. The simulation study also showed that the signal priority component was effective in assigning priority to the equipped trucks and buses based on a pre-defined hierarchy of control [6]. The MMITSS system development is now in Phase III in which the goal is to make deployment readiness enhancements to the MMITSS prototypes that were developed and field tested in Phase II.

2.2 UDOT Implementation of MMITSS

From 2015, Utah DOT decided to initiate efforts to deploy (vehicle-to infrastructure) V2I systems. The goals of initial deployment included: (1) to gain hands-on experience with the procurement and installation of DSRC equipment; (2) to deploy an application that could yield a tangible benefit (to justify the cost of installation); and (3) to equip a corridor that could subsequently be used for the development, testing, and implementation of other connected vehicle applications, including those that will ultimately be installed in vehicles. In partnership with the Utah Transit Authority (UTA), UDOT decided to deploy the MMITSS signal priority application on transit vehicles. The objective was to improve the reliability of bus service while optimizing the use of available green time [7].

Roadside DSRC radios have been installed at 30 intersections in Salt Lake City, UT. When the buses are behind schedule, the system grants priority so they can get back on schedule. The project developed a schedule-checking module and built upon the MMITSS platform. Preliminary operational results showed that transit reliability on this corridor for bus route #217 has improved from 86% to 94% [7].

UDOT MMITSS deployment project is one of the first DSRC systems in regular operation in the United States. In addition, this project is one of the first completed projects that address the signal phase and timing (SPaT) challenge. The SPaT challenge is organized by the American Association of State Highway Transportation Officials (AASHTO), the Institute of Traffic Engineers (ITE), and ITS America (ITSA) through the Vehicle to Infrastructure Deployment Coalition (V2I DC) that have challenged state and local public sector transportation infrastructure owners and operators to work together to achieve deployment of roadside DSRC equipped units to broadcast SPaT in real-time at signalized intersections on at least one road corridor or street network (approximately 20 signalized intersections) in each of the 50 states by January 2020. The main mid-term goal of the SPaT Challenge is to deploy DSRC broadcasts of the SPaT messages, and the long-term objective is to sustain the operation of connected vehicle applications that utilize the SPaT messages [8].

3 Ongoing Projects

3.1 CV Pilots

To spur the early deployment of CV technology, in September 2015 the United States Department of Transportation (USDOT) awarded three (3) cooperative agreements to Wyoming Department of Transportation, New York City Department of Transportation (NYCDOT) and Tampa Hillsborough Expressway Authority (THEA). Two of the three sites, NYCDOT and THEA, are deploying CV technology to improve traffic signal progression along their respective corridors [9].

NYCDOT is deploying two applications that deal with interactions of CV and traffic signal systems. Pedestrian in Signalized Crosswalk application will use pedestrian detection technology to indicate the presence of a pedestrian in a crosswalk at a signalized intersection. Pedestrian's presence will be detected by pedestrian detection equipment as a pedestrian passes through the crosswalk and notify the vehicles of a pedestrian's presence. The MMITSS PED-SIG application is aimed at supporting visually impaired to cross the street. The application will be implemented on a portable mobile device communicating in both cellular and DSRC spectrum and able to request service at signalized intersections utilizing the PED actuation operation. For the intelligent traffic signal system NYCDOT is proposing to use CV data as an input to the existing Adaptive Control Decision Support System by augmenting data from the toll tag reader system that is used to provide travel time and speed information.

Tampa Hillsborough Expressway Authority (THEA) is deploying two applications developed under the auspices of MMITSS at various locations within the pilot deployment area. I-SIG is being deployed along two corridors to improve traffic progression and increase travel time reliability. TSP component of MMITSS is also being deployed along four routes used by Hillsborough Area Transit Authority.

3.2 Intelligent Real-Time Isolated Intersection Traffic Control System (IICS)

This study is led by University of Florida under several projects funded by National Science Foundation (NSF) and Florida Department of Transportation (FDOT). The research foundation of this project is to optimize traffic signal control and vehicle trajectories simultaneously at isolated intersections [10], in a mixed traffic and low demand condition, where AV, CV and conventional vehicles co-exist [11].

The algorithm takes arrival information of vehicles when they enter the communication range as the input and then calculates optimal vehicle trajectories for a given signal timing plan. All vehicles are supposed to accelerate to the maximum allowed speed and keep the saturation flow rate to utilize the green time to its maximum efficiency. The signal timing parameters are enumerated and the optimal signal timing is selected which minimizes the average travel time delay (ATTD). A rolling horizon scheme is applied to conduct the optimization over the time horizon for new approaching vehicles [10].

The proposed IICS system was implemented and tested at the Traffic Engineering and Research Laboratory (TERL), a FDOT closed-course facility. A total of six vehicles participated in the field demonstration, including four CVs (one AV with SAE level four and DSRC connectivity), and two conventional vehicles. The AV is a hybrid Toyota Highlander equipped with different types of sensors including GPS, IMU and Lidar sensors for localization and obstacle avoidance. The intersection is equipped with a Doppler-based radar sensor system to detect and classify all approaching vehicles, which provides arrival information for conventional vehicles. Econolite Cobalt controller is used to control the traffic signal. A NTCIP 1202 protocol application is developed to control the controller. Both CVs and the intersection have either Onboard Units (OBUs) or Roadside Unit (RSU) for transmitting and receiving DSRC messages. Results showed that the system was able to provide optimal trajectories to both CVs and AVs in a very short time, which reduced delays due to unnecessary stops. However, it was difficult for the CV drivers to follow the suggested trajectory closely. The root mean square errors (RMSE) between planned trajectories and actual trajectories range from 4.3 ft to 64.9 ft [12].

3.3 Traffic Optimization for Signalized Corridor (TOSCo) Project

The TOSCo project is sponsored by Federal Highway Administration (FHWA) and performed by the V2I Consortium of CAMP LLC, in conjunction with three university partners, including University of Michigan Transportation Research Institute (UMTRI), Texas A&M transportation research institute (TTI) and University of California at Riverside (UCR). This project aims at optimizing traffic flow and minimizing vehicle fuel consumptions and emissions at signalized arterial roadways. The system applies vehicle-to-vehicle (V2V) communications to form strings to reduce headways and vehicle-to-infrastructure (V2I) communications to estimate real-time queue length at intersections. An eco-trajectory planning algorithm is designed to smooth trajectory and reduce number of stops.

TOSCo equipped intersections are constantly broadcasting information about SPaT, MAP, and Roadside Safety Messages (RSM). The RSM includes predicted queue length and green window information, which provides a feasible time interval for TOSCo equipped vehicles to pass the intersection. When TOSCo-equipped vehicles are outside the communication range, the vehicles would operate in a free-flow mode. They would operate under either manual control, adaptive cruise control (ACC) or Cooperative Adaptive Cruise Control (CACC), depending on its position in the string and whether it is following another connected vehicle (CV). When a TOSCo-equipped vehicle enters the communication range of the intersection and receives SPaT, MAP and RSM, the vehicle would then plan a speed trajectory that allows it to either pass through the intersection without stopping (either by speeding up slightly, maintaining a constant speed, or slowing down slightly) or to stop in a smooth, coordinated fashion that would minimize the amount of time stopped at the intersection. TOSCo-equipped vehicles that have to stop at an intersection would perform a coordinated launch maneuver at the start of the signal phase that would allow them to clear the intersection in a more efficient manner, which greatly increases the intersection capacity. Planning the appropriate trajectory requires information from the infrastructure (i.e., SPaT and RSM). To provide accurate queue length and green window information, a set of infrastructure algorithms are developed, depending on different data collection and sensing technologies (e.g., CV trajectory data, loop-detector data and radar data).

A simulation platform that integrates a vehicle behavior module, an infrastructure component and a fuel and emission evaluation module is designed and implemented in VISSIM. A low speed corridor (Plymouth Rd, MI) and a high speed corridor (SH105, TX) are modeled in VISSIM and calibrated with real-world data. Different scenarios with varying TOSCo-equipped vehicle penetration rates are simulated. Simulation results demonstrated that TOSCo functions improve mobility in terms of total delay, stop delay, number stops, average speed and total travel time. Meanwhile, TOSCo functions also demonstrated smoother traffic flow especially at higher penetration rates, which has a significant positive impact on discharging queues and maintaining motion which improves fuel economy and reduce emissions.

4 Potential Future Projects

Beside the completed and ongoing projects, there are studies that have the potential for future development. In this section, selected projects are briefly introduced.

The queue spillback problem in freeway on-ramp meters that are adjacent to traffic signals is investigated by Kan et al. [13]. During peak hours, the traffic signals employ long signal cycles to maximize intersection capacity which results in long platoons of freeway-bound traffic advancing toward the on-ramp within a short period, which quickly fills up the on-ramp and causes spillback. Currently, the problem is addressed by suspending ramp metering or relaxing the metering rates that will inherently increase freeway delay. It was shown that V2I communication allows the traffic signals to adjust cycle length considering on-ramp queue length. Improved arterial signal timing reduces both arterial and freeway delay.

Honda demonstrated a smart intersection concept in Marysville, Ohio, that utilizes high-resolution cameras to track vehicles, pedestrians and cyclists' trajectories in the vicinity of the intersection and generates proxy messages to provide warning messages to connected vehicles. Wireless communication systems, computer vision recognition systems, infrastructure control systems, and in-vehicle telematics and alert systems are all connected and coordinated [14]. This infrastructure based sensing and communication platform has great potentials in both traffic control and facilitating the automated driving system (ADS).

A decentralized energy-optimal control framework for signal-free urban intersection problem using CAV technology is addressed by Malikopoulos et al. [15] The goal of this analytical solution is to minimize the fuel consumption subject to a throughput maximization requirement. The study concluded that feasible solutions satisfying all safety constraints always exist. The effectiveness of the framework was illustrated through simulation. The results show CAVs can conserve momentum and fuel while their travel time can be improved by 20%.

5 Conclusion

In this chapter, a brief review of the completed and ongoing research projects that presented during the AVS 2018 are discussed. The review includes the specifications of the projects and the challenges of the new technology. In addition, a few interesting research efforts which may need to field implementation in the near future are also introduced.

Besides existing research and deployment projects, the following bullet points summarize the research needs from the community:

- Trajectory-based signal control systems have a lot of potentials to be investigated.
- With low CAV penetration rate, other data sources such as shared mobility companies data should be integrated.
- Sharing algorithms source code and creating an open source community between researchers can be very helpful in speeding up the future development.
- Pedestrians should not be left out in evaluation of advanced CAV intersection control systems.
- There is a need for a generic simulation platform that can precisely replicate the CAV behaviors, V2X messaging, and interactions with human-driven vehicles.
- The benefits of CAV technology on non-signalized intersections should also be studied.
- It is necessary to rethink about intersection capacity analysis when CAV penetration rate increases.

References

1. University of Arizona, University of California PATH Program, Savari Networks, Inc., Econolite, MMITSS – Phase II: System Development, Deployment and Field Test – Final Report (2016). http://www.cts.virginia.edu/wp-content/uploads/2014/04/53-MMITSS-Phase-2-Final-Report-FINAL-092520161.pdf
2. Feng, Y., Head, K.L., Khoshmagham, S., Zamanipour, M.: A real-time adaptive signal control in a connected vehicle environment. Transp. Res. Part C: Emerg. Technol. **55**, 460–473 (2015)
3. Zamanipour, M., Head, K.L., Feng, Y., Khoshmagham, S.: Efficient priority control model for multimodal traffic signals. Transp. Res. Rec.: J. Transp. Res. Board **2557**, 86–99 (2016)
4. Beak, B., Zamanipour, M., Head, K.L., Leonard, B.: Peer-to-peer priority signal control strategy in a connected vehicle environment. Transp. Res. Rec.: J. Transp. Res. Board **2672**, 15–26 (2018)
5. Ahn, K., Rakha, H.A., Kang, K., Vadakpat, G.: Multimodal intelligent traffic signal system simulation model development and assessment. Transp. Res. Rec.: J. Transp. Res. Board **2558**, 92–102 (2016)
6. Zamanipour, M., Head, L., Ding, J.: Priority system for multimodal traffic signal control. In: Presented at the Transportation Research Board 93rd Annual Meeting (2014)
7. Leonard, B.: Strong signal; Utah breaks new ground with connected vehicles. https://www.roadsbridges.com/strong-signal
8. SPaT challenge overview (2018). https://transportationops.org/spatchallenge

9. Hartman, K.: Connected Vehicle Pilot Deployment Program. U.S. Department of Transportation (2018). https://its.dot.gov/pilots/
10. Li, Z., Elefteriadou, L., Ranka, S.: Signal control optimization for automated vehicles at isolated signalized intersections. Transp. Res. Part C: Emerg. Technol. **49**, 1–18 (2014)
11. Pourmehrab, M., Elefteriadou, L., Ranka, S., Martin-Gasulla, M.: Optimizing signalized intersections performance under conventional and automated vehicles traffic. arXiv e-prints arXiv:1707.01748 (2017)
12. Omidvar, A., Pourmehrab, M., Emami, P., Kiriazes, R., Esposito, J.C., Letter, C., Elefteriadou, L., Crane III, C.D., Ranka, S.: Deployment and testing of optimized autonomous and connected vehicle trajectories at a closed-course signalized intersection. Transp. Res. Rec.: J. Transp. Res. Board **2672**(19), 45–54 (2018)
13. Quain, J.R.: With cameras that know dogs from Dodges, Honda is making intersections safer, Digital Trends (2018). https://www.digitaltrends.com/cool-tech/honda-smart-intersection-marysville/
14. Kan, X., Lu, X., Skabardonis, A.: Increasing freeway capacity by efficiently timing its nearby arterial traffic signals. Transp. Res. Rec.: J. Transp. Res. Board **2672**(18), 27–34 (2018)
15. Malikopoulos, A., Cassandras, C., Zhang, Y.A.: Decentralized energy-optimal control framework for connected automated vehicles at signal-free intersections, automatica. arXiv: 1602.03786 (2017)

Truck Platooning: Connectivity Enabled, Grounded in Safety, Properly Tested

Joshua P. Switkes(✉), Rod McLane, Shad Laws, and Mark Luckevich

Peloton Technology, Inc., 1060 La Avenida, Mountain View, USA
{Josh, Rod, Shad, Mark}@peloton-tech.com

Abstract. Development of a truck platooning system that is effective and safe requires discipline along the entire process. This includes setting the right goals for performance, designing the right functionality, implementing that functionality, and thorough testing.

Keywords: Platooning · Safety · Commercial vehicles

1 Truck Safety Today

Trucks today are remarkably safe, but they still cause an unacceptable number of collisions and fatalities. Although trucks are involved in far fewer collisions than passenger vehicles on a per-mile basis, truck collisions tend to be more severe and more noticeable. In 2016, for example, there were 475,000 police-reported crashes involving large trucks, which amounts to 1.65 crashes per million large-truck miles traveled. Approximately 0.8% of these crashes were fatal[1].

When a truck is involved in a fatal collision, it's the people in the nearby vehicles, rather than the truck driver, who are most often harmed. So, in a very real sense, truck safety matters to everyone on the road. Recent improvements to truck technologies, like mandatory anti-lock braking and the Collision Mitigation Systems (CMS) now offered by all truck makers (OEMs), have been shown to dramatically reduce collision rates. A Department of Transportation/Volvo Trucks study has found that tractor-trailers equipped with only a collision warning system, or a collision warning system in combination with adaptive cruise control and electronic stability control, were involved in 37% fewer situations that had potential to result in a rear-end collision. A Con-way study found a 71% reduction in rear-end collisions for tractor-trailers equipped with collision-avoidance systems, including automatic emergency braking, electronic stability control, and lane-departure warning[2].

The Peloton platooning system is built on top of and alongside these active safety systems.

[1] Large Truck and Bus Crash Facts 2016. Federal Motor Carrier Safety Administration Analysis Division (May 2018). Pages 17 and 55.

[2] The Use of Forward Collision Avoidance Systems to Prevent and Mitigate Rear-End Crashes. National Transportation Safety Board (2015). Pages 18–19.

G. Meyer and S. Beiker (Eds.): AUVSI 2019, LNMOB, pp. 92–99, 2019.
https://doi.org/10.1007/978-3-030-22933-7_10

2 The Peloton Platooning System Safety Principles

The development of the Peloton platooning system includes the following nine components listed below. We will dive into a few of these in greater detail throughout this chapter:

- **Start with Industry-Leading Technology**: We build truck platooning on top of the leading active safety technologies available for trucks.
- **Supervise Platooning**: We actively constrain the Operational Design Domain (ODD) to road and driving conditions for which we have developed and validated.
- **Compare to an Appropriate Benchmark**: We analyze and compare safety statistics with the following in mind: How does the risk change when the driver presses the platooning button?
- **Implement the Right Functionality**: We implement the right functionalities to achieve our safety goals in mixed traffic/conditions on real vehicles.
- **Implement the Functionality Right**: We follow ISO 26262, the standard for functional safety of electrical and electronic systems in production automobiles, as defined by the International Organization for Standardization (ISO) in 2011. It is our structure for developing and validating a safe system.
- **Manage Variation in Vehicle Spec and Condition**: Real-world vehicles and road conditions vary considerably, and we design with this in mind, at the core of our engineering analysis
- **Keep the Driver at the Center**: The driver is the key part of the safety of Peloton platooning, and we design and test with the driver's safety and comfort front and center.
- **Collaborate with Industry and Government**: Peloton platooning is the result of joint development with truck OEMs and Tier-1 suppliers and deploys collaboratively with both state and federal government.
- **Test Properly**: We make a tremendous effort to ensure that the testing itself is safe, and that the safety testing is thorough and appropriate for our systems and products.

2.1 Guided by an Appropriate Benchmark

We strive to increase the safety of trucks on the road. Our goal is for drivers to be safer after they push the platooning button.

To do this, we compare platooning vehicles not just to average vehicles, but to the vehicles that lead the industry in safety: trucks equipped with active safety systems (Fig. 1).

This establishes a benchmark and guides us in our iterative system-design process to improve upon the current safety state of the art.

2.2 Implement the Right Functionality

2.2.1 Connected Braking

Peloton's direct V2V communication, based on the industry-standard DSRC, allows two platooning trucks to accelerate and brake together as a single system. Specifically,

Fig. 1. Platooning engage button

the DSRC connection enables the follow truck in a platoon to react nearly instantaneously to acceleration and braking by the lead truck.

In communicating the lead-truck braking force to the follow-truck system, following distance is reduced, perception delay is eliminated, and reaction delay – from the typical braking sequence of perception/reaction/braking – is vastly shortened. This is a key differentiator between Peloton platooning-equipped trucks and manual or radar-based braking.

The platooning system reacts extremely quickly, accurately, and reliably. Its reaction time is quicker than that of the human driver. It applies the brakes precisely to match or exceed the braking by the lead truck. And, unlike a human driver who may sneeze, sip coffee, or even text, it doesn't become distracted.

2.2.2 Platoon Proximity Dissolve

The platooning system has the capability to dissolve the platoon when other vehicles are in close proximity to the front of either vehicle. Through integration with the CMS, we utilize radar and camera sensor data to monitor traffic in front of the lead truck. V2V is used to relay this sensor information to the rear vehicle. The system in the rear vehicle uses the sensor data to evaluate traffic, or "targets" in front of the lead truck. If events occur in close proximity to the lead truck (e.g., a car cutting closely in front of the lead truck), then the system will preemptively dissolve the platoon. In this way the system separates the trucks to a safe distance and slows the follow truck relative to the lead vehicle before situations occur that may require hard braking.

2.2.3 Platoon Dissolve

While we do not rely on the drivers to assess conditions like V2V connectivity, we do train drivers to dissolve the platoon themselves when upcoming conditions may be inappropriate.

When a driver decides to end platooning, the follow-truck system will increase the gap between the trucks until a safe manual follow distance is reached. At that point, the follow driver takes control using the brake or accelerator pedal. During dissolve, the system will respond to the lead truck's braking or to avoid a vehicle that cuts in between the two trucks.

2.2.4 Cut-in Detection and Reaction

While the close proximity of the two platooning trucks makes it easier for other vehicles to navigate ahead or behind the platooning trucks, another vehicle may cut between them. Platooning drivers likely will recognize a cut-in threat before the system does, and therefore drivers are trained to dissolve the platoon to make room for the cut-in. But if they don't, the system will initiate dissolve immediately upon detection and simultaneously alert the drivers.

2.2.5 Driver Awareness Video and Info Display

In platooning, the follow driver is as aware and engaged as the lead driver in the driving task. While follow drivers have a generally clear view of the road, the platooning system provides a video feed from a forward-facing camera in the lead truck. This enables them to see vehicles, objects, and road features like entry ramps ahead of the lead truck. Follow drivers look at the video view as part of their mirror-scan cycle.

2.2.6 Driver Teamwork Through Voice Communications

Platooning has shown a tremendous potential to facilitate teamwork between drivers when voice communications are enabled between the platooning drivers. Drivers use a hands-free driver-to-driver radio communication feature to meet on the road, share information about road conditions and coordinate maneuvers such as lane changes.

2.2.7 Cybersecurity: V2C and V2V and IntraCloud

A safe platooning system requires security in several key areas, including at the device and corporate levels and in the cloud infrastructure. We continually evaluate and adopt recommended security practices to ensure that our system remains tightly secured.

2.3 Implementing the Functionality Right

Even the best planned functionality is only as good as its implementation, so Peloton leverages leading development practices from both the automotive and high-tech industries in our joint development programs with OEMs and Tier-1 suppliers. The result is a high-reliability system that can also rapidly add new capabilities without requiring a multi-year validation cycle.

2.3.1 ISO26262

Together with our development partners, Peloton has been guided by ISO 26262, the leading standard for functional safety. Following this standard is invaluable to several critical aspects of our development process:

- Traceability: The International Organization for Standards (ISO) process allows us to trace back to high level safety goals every vehicle test, every Hardware in the Loop test, and every Software in the Loop test.
- Full coverage: The ISO process gives confidence that our analysis and testing cover key safety areas.

- Rapid re-validation: ISO gives us the ability to rapidly re-validate through our test processes on Hardware (HW) in the Loop, Software (SW) in the Loop, and track testing.

2.3.2 HW Implementation

The Peloton Electronic Control Unit (ECU) is architected to meet our safety goals. Guided by ISO 26262, the ECU and its key components have undergone ASIL assessment and validation.

2.3.3 SW Implementation

On the architected hardware, Peloton has implemented a multi-tiered software structure. We have taken leading Operating Systems (chosen for each processor) and implemented a safety critical middleware. We have found this combination of software has the iteration capability of the best prototype-level middleware, combined with full traceability and other safety requirements from production operating systems. This allows rapid development while meeting our safety goals.

2.4 Manage Variation in Vehicle Spec and Condition

Peloton is working towards a significant reduction in the collision rate compared to trucks on the road with advanced safety systems. To do this, we actively manage the variations in truck specifications and equipment conditions, and we have extensively tested the braking capabilities of tractors and trailers under a wide variety of vehicle and road conditions.

2.4.1 Stopping Distance Variation

Today's trucks vary considerably in on-board equipment, like brakes or wheelbases. And the condition of their on-board equipment varies from truck to truck. In addition, trucks carry different types of cargo, which affects their total mass, and, consequently, their stopping distance. For example, one truck might be able to stop 100 ft or more sooner than another.

This is one of the primary reasons why truck drivers are instructed to follow at quite large following distance such as six or seven seconds of time headway. In the real world, however, drivers often follow at what are considered unsafe distances.

To make the road safer, Peloton platooning uses knowledge of the two trucks to determine whether they should platoon. Using sensor data including stopping-distance variation, the system selects which trucks are allowed to platoon safely, and which trucks in a team should lead or follow.

2.4.2 Vehicle and Equipment Variation Assessment

We generally separate areas of variation into two categories: Those we can measure or determine on the vehicle, and those we cannot.

Examples of the ones we can measure or determine are the vehicle mass (we calculate that in real time), or the type of brakes on the tractor. Ones we cannot measure in real time today include tire-tread depth, or the height of cargo inside the trailer.

Therefore, our platooning gap is set to be large enough to absorb uncertainties in stopping distance due to these unknown factors. As we continue to improve our ability to measure more of these factors in real time, we may safely decrease the platooning gap and improve the overall efficiency of the system.

2.4.3 Overall Improvement Through Connected Driving

When the functionality is combined with our design- and test-methodology, the end result is that we can operate the trucks much closer together than what, in the past, has been considered safe.

MANUAL DRIVING

Let's look at manual driving – that is, typical, individual trucks on the road. When a truck brakes, there is a lag before it starts to slow down. For drivers following that truck, it can take 1.0 to 1.5 s to perceive and react to the braking of the lead truck. When following drivers apply their brakes, their own trucks experience a brake lag. In addition, there could potentially be a big difference in braking distance – 100 ft or more – compared to the lead truck.

Combining these factors, the safe following distance for a manually driven truck is many hundreds of feet, but because drivers feel it is impractical in many cases to follow a such distances, collisions occur too frequently for heavy trucks, and many collisions occur at a high velocity.

FOLLOWING WITH CMS and ACC

Because sensors do not get distracted the way human drivers do, a vehicle equipped with an Automatic Emergency Braking and/or a collision mitigation system, can eliminate the attention lag time for braking. They also can dramatically reduce the perception distance, and partially reduce the reaction distance. Trucks equipped with such systems can reduce the appropriate following distance to about 300 ft, while lowering collision rates. They also reduce the impact velocity of many collisions, reducing the severity.

In the real world, however, such following distances are difficult to maintain. Other vehicles cut in between these following vehicles frequently, lowering the real-world following distance to 150–200 ft, creating higher risk. Neither of these systems help at all with braking differences between vehicles, or with brake lag.

PLATOONING

When platooning, the V2V and V2Cloud dramatically improve several of these factors. The perception and reaction times are reduced dramatically, and because the follow truck can react to the lead truck brake application (before the lag), the lag is effectively eliminated (the two trucks experience the lag at nearly the same time).

The Peloton platooning system also has partial information about the braking capability of the lead truck, which eliminates a large portion of the uncertainty in braking between the two vehicles.

The end result is a dramatically shorter following distance than that prescribed for manual driving. In addition, the residual collisions that do occur are mostly at very low velocities, and thus are far less severe than a typical collision.

2.5 Keep the Driver at the Center

In the Peloton platooning system, drivers are in primary control of their trucks. To ensure that the system works well for them, we are guided by these principles:

- Be sure the driver is aware of the actions and state of the system
- Provide the driver with enough information to know what actions to take or not take
- Keep the driver engaged in the driving task

2.6 Test Properly

With the right functionality and the right intended implementation, it is then critical to conduct appropriate testing to be sure the implementation meets the requirements.

As is common in the industry, the testing of the system is a multifaceted activity using a number of tools and test methods. Fundamental to testing properly is the design of the test methods and test plans. These are derived and made traceable to the system requirements and the system safety goals, as guided by ISO26262. Testing to the requirements is commonly known as verification and validation of the system or V&V testing. In this testing, one validates that the design meets the requirements and verifies that the system performs as designed.

Peloton has embraced industry standard test methods and developed the tools and test methods to fully validate the system. This includes a Software in the Loop tester (SIL) and Hardware in the Loop tester (HIL), vehicle on-track test methods and procedures, and an extensive test process to verify that the system is ready for on highway use.

2.6.1 ISO26262 and Testing

ISO26262 provides an excellent framework for the testing, allowing each requirement to be tested in one or more tests. The ISO process starts with a Hazard and Risk analysis (HARA) which is used to determine the safety goals of the system. These in turn are used to derive functional and technical safety requirements which then provide the framework for testing the safety of the system. The tests derived from the analysis se can be vehicle level track tests, they can be Hardware in the Loop test, or Software in the Loop. Using this test process, all of the tests are traceable back to the safety goals so that we can gain confidence that the function has been tested and the safety goal has been met.

2.6.2 HIL Capability

The Peloton platooning system is heavily tested in a HIL test environment. This consists of Peloton platooning hardware being exercised in a variety of ways to mimic real-world use of the system. This includes a broad set of tests following the ISO26262 process. In addition, we can take data collected from real world driving conditions, feed the data into the HIL tester, and repeatedly test a variety of operational conditions.

This HIL tester is very importantly made from actual Peloton platooning hardware, to make sure it is properly testing any detail issues like timing of signals or other issues that might not be found in a software-only tester.

2.6.3 Stress Testing

Peloton understands that the real world can be unpredictable and that we must make every effort to test the robustness of the system. Similar to the ISO2622 testing, this can be accomplished with SIL, HIL and vehicle track testing. Peloton is continually deriving tests to feed into our test suite to exercise the system in as many ways as possible. In addition, after all our safety goals were met, we added testing on highway with trained test drivers to get the real-world experience.

2.6.4 Testing Safely

Safe testing really includes two aspects: safety of testing and testing for safety. Before Peloton software ever goes on the road, it goes through an extensive process of peer code review, unit tests, SIL and HIL testing, and then a gamut of track tests that test a safety-critical subset of the full set of functional tests derived from ISO26262.

Even then, it is only driven by highly trained test drivers. These drivers are not only highly trained truck drivers, they have also been trained in various evasive maneuvers and other practices to enhance safety during pre-production testing. Only after production validation (including HIL, SIL, and mileage accumulation) has been completed, can a non-test driver drive with the system. Peloton also doesn't disable the existing active safety systems on test vehicles. In fact, these systems are a core part of the Peloton platooning system and remain fully active with the system onboard.

3 Conclusion

Since its inception, Peloton has had the singular focus of creating and developing products that improve the safety for the trucking and transportation industry while providing real world improvements in operational efficiency. Following the mantra that "Safety is our North Star", we have created a culture of transparency to make sure our employees, customers, partners, and the general public understand our safety approach.

This chapter summarizes the nine components of our safety approach that have guided the development of the Peloton Driver Assistive Truck Platooning product, which we feel is the foundation for commercial vehicle automation. As this technology is commercially deployed, we along with our customers and partners are excited to see the benefits in safety and efficiency this technology will deliver.

Automation and Adverse Weather

David Neumeister(✉) and Douglas Pape(✉)

Battelle, 505 King Avenue, Columbus, OH, USA
{neumeister, pape}@battelle.org

Abstract. Like other vehicles, vehicles with automated driving systems have to handle a variety of adverse weather conditions. Furthermore, when anticipated weather for all or part of a trip is outside the operational design domain of the system, the vehicle itself or the humans responsible for it need to take appropriate action.

This chapter opens with a review of literature on adverse weather. Limited experiments with real and simulated adverse weather exemplified current systems' capabilities and gaps in performance. Results were presented at two meetings to solicit broad input on how all stakeholders can cooperate to improve the performance of automated driving systems in adverse weather.

Keywords: Vehicle automation · Road weather · Adverse weather

1 Background

Of the more than 5,748,000 annual motor vehicle crashes in the United States, 22% are related to weather (FHWA 2017). Current vehicles with automation features assist with steering and braking. They are not designed to operate in all weather conditions. The ability of automation systems can be affected by the weather (atmospheric conditions) and road weather (surface conditions).

A human driver of a conventional vehicle needs information on weather and its effects. If current conditions are dark or rainy, the driver will turn on the headlights. If the road surface is suspected to be slippery, the human will drive more slowly and avoid sudden maneuvers. If a severe storm is predicted to arrive before a planned trip can be completed, the driver may elect not to make the trip at all.

All of these same scenarios apply to automated vehicles, but they are complicated by the capabilities of the driving automation system. SAE Recommended Practice J3016 (SAE International 2018) relies on the concept of an operational design domain (ODD), which is the set of conditions under which a driving automation system is designed to function, including geography, roadway, and weather. The human (or the vehicle itself) must decide whether to begin or continue a trip using a driving automation system. Therefore, there is a need for information on current and forecast conditions for the planned route.

G. Meyer and S. Beiker (Eds.): AUVSI 2019, LNMOB, pp. 100–107, 2019.
https://doi.org/10.1007/978-3-030-22933-7_11

2 Literature Review

Disseminating weather information is not new, and the tools of modern technology are being applied. However, automation in adverse weather raises new questions in human factors.

2.1 Common Sensors for Automation Technology

Manufacturers of current automated vehicles use a combination of video cameras and radar to sense the other vehicles, objects, and lane markings. Some vehicles in development also use light detection and ranging (LiDAR), which allows 360-degree vision of a three-dimensional (3D) scene. LiDAR is limited in weather. For more than 20 years radar has been used on trucks to estimate the range to the leading vehicle, for automatic cruise control or, more recently, automatic emergency braking. Radar waves are much longer than the infrared waves used for LiDAR, so they can travel better through airborne precipitation but their resolution is not as fine.

2.2 Existing Systems for Road Weather Information

The Federal Highway Administration (FHWA) has a number of initiatives for disseminating road weather information.

Pathfinder. This program provides travelers with a consistent assessment of the weather's effects on travel. Supported by the FHWA and the National Oceanic and the National Weather Service, Pathfinder facilitates cooperation between public and private forecasters.

Road Weather Performance Management (RW-PM). Pikalert® incorporates vehicle-based measurements of the road and surrounding atmosphere with other, more traditional weather data sources for use in this connected vehicle application. The processed data are pushed to travelers in near real time and through web-based user interfaces and as in-vehicle advisories. Information in the vehicle can, in principle, be read by the vehicle itself as well as by the human driver.

Weather Data Environment (WxDE). This research tool collects and shares transportation-related weather data with a focus on connected vehicle applications. Weather data could be provided directly to vehicles or to a dispatcher responsible for a fleet of automated vehicles.

ARC-IT. The Architecture Reference for Cooperative and Intelligent Transportation guides organizations in packaging and distributing information on, for example, icy road conditions, high winds, and dense fog.

Requirements for Vehicle-to-Infrastructure (V2I) Weather Applications. An application being considered by the J2945/3 technical committee of SAE International is a road weather module for automated vehicles (SAE International 2012). Part of a larger effort to expand the types of messages communicated through connected vehicle systems, this standard addresses the road weather needs of various road users.

2.3 Vehicle Sensors to Aid Weather Prediction

The sensors used to support automated vehicle operations may be suited to help other vehicles and agencies as well. "Ground truth" conditions detected by several automated vehicles can provide alerts to vehicles approaching a geographic area, agencies maintaining the roadways, and even agencies providing weather services. Information could be based on some combination of automated vehicle sensor failures (e.g., due to road snow or ice) and, perhaps, on-board vehicle sensors (e.g., wiper status) (USDOT 2016).

2.4 Human Factors of Automation in Adverse Weather

Different types of adverse weather create different challenges to which automated vehicles and drivers must respond (Sundararajan and Zohdy 2016). A significant amount of research surrounding human performance on similar tasks can offer concrete guidance on how human operators are likely to respond to automated systems (Hergeth et al. 2017; Schwarz et al. 2016).

One factor that is highly correlated with driver engagement with the system, and that has a significant effect on how drivers interact with the system, is trust (Koglbauer et al. 2017). Driver trust in an automated system varies depending on a number of factors: how accurate their understanding of system operation is, their previous experience with the system or similar systems, and their willingness to give up control of the vehicle (Hergeth et al. 2017; Schwarz et al. 2016). Drivers can be either over-trusting or under-trusting of an automated system, which both present human factors challenges.

Strong winds may not be obvious to a human not steering the vehicle, especially on road surrounded by flat terrain or on long bridges. High winds introduce new forces that complicate steering. If the vehicle does not warn the driver that it is compensating for high winds, then the driver may not compensate adequately in the transition to manual control (Hergeth et al. 2017; Schwarz et al. 2016).

During a critical situation, such as one caused by adverse weather, it is incumbent upon the driver to make crucial decisions in a limited time in order to preserve safety. If the driver is not ready to re-engage or is distracted, the driver may not have enough time or cognitive capacity to respond to an emergency situation (Louw et al. 2015).

3 Experiments

Three vehicle models commercially available in the United States were used for these tests. All three had machine vision (video cameras), and two had a radar. The tests challenged perception systems across a variety of simulated adverse weather conditions in a controlled outdoor setting. Tests were performed at the Transportation Research Center (TRC) in East Liberty, Ohio. Vehicles were tested in a baseline condition and with iced sensors, in falling rain, and on a wet road. Limited tests with falling snow and sun glare were also run (Fig. 1).

Fig. 1. The vehicle used as the lead vehicle drives through simulated rain on the Winding Road Course

3.1 Results

The rain and the water on the pavement did not appreciably affect the ability of the vehicles to recognize the lane lines. In nearly every test run, warnings of a lane departure were appropriately given and restoring steer input was applied. A surprisingly small amount of blown snow covering one lane line prevented one of the vehicles from tracking the line.

The falling rain affected performance of high- and low-speed following on several occasions, but the effect was brief interruptions from which the vehicles quickly recovered. In some cases, the cloud of water prevented the vehicle from perceiving the vehicle ahead, causing the vehicle to speed up; in others, the vehicle perceived a heavy cloud of water as an obstacle and braked. One instance of such braking was strong enough that it would have been uncomfortable to passengers.

Ice on the sensors stymied all automation systems. The initial 6-mm thickness of ice was too much for radar and vision systems. Ice was scraped off in increments. None of the vehicles could recognize lanes with any amount of ice over the cameras. Lane-keeping was possible in two vehicles when ice was completely removed from the windshield; on the third the windshield also needed to be cleared of the residual water. One vehicle was able to perform lane-keeping and low-speed following with an iced-over radar sensor and a cleared windshield camera. Otherwise, none of the vehicles could successfully perform following with the radar even thinly iced over.

Lane Departure on a Straight Roadway. All three vehicles needed some time after first seeing lane lines (several seconds at 45 mph or 72 kph) to recognize them. All vehicles could engage with one lane marking and a longitudinal seam in the road. The vehicles were generally able to detect lane lines on straight roads, in both the baseline and rain test runs.

High-Speed Following. All three systems intermittently lost tracking in heavy rainfall. Two were able to quickly regain tracking and to maintain a safe following distance at all times. The third vehicle disengaged automatic cruise control in heavy rain and applied hard braking, so it failed safe.

Traffic Jam Assist. The speed in these tests ranged from a full stop to 12 mph (19 kph). Two vehicles had a greater minimum following distance in the rain than in the clear weather; the third followed more closely in the rain.

3.2 Human Factors Observations

The three models had different human-machine interfaces. All vehicles indicated the driver-selected following distance for cruise control graphically, one differently than the others. They differed in how they alerted the driver of a potential forward collision —using different combinations of text, shape change, color change, and sound. One vehicle audibly confirmed successful engagement of automated functions. The other two vehicles audibly alerted the driver only when a function disengaged or a warning was issued, though one gave no indication when it lost lane tracking.

4 Stakeholder Engagement

Input from a broad range of stakeholders was solicited at two meetings in 2018. Participants at the Transportation Research Board (TRB) meeting came principally from state and local transportation agencies. Industry was better represented at the discussion during the Automated Vehicles Symposium. The gaps that were identified and the recommendations that were voiced will guide future work.

4.1 Roles to Support Automated Vehicles Are Unclear

SAE J3016 is clear that the automation level of a driving automation feature is assigned (by the manufacturer) rather than measured (by a testing agency). The operational design domain of a feature, the set of conditions under which it is designed to function, is set by the vehicle manufacturer or system supplier.

What is not so clear is who has the responsibility to determine whether current or forecast conditions are within a system's operational design domain. At different levels of automation, a human may request control from the vehicle, or the vehicle may request that the human resume the dynamic driving task. When a trip is being contemplated, one of the decisions is whether the vehicle, any driving automation features, and the human are capable of handling the forecast conditions. Conceptually, a driver may choose to forego a trip, a vehicle may refuse to initiate a trip, or a remote dispatcher may deny the trip.

State and local operating agencies are concerned that they may be asked to take on a new responsibility that they are not prepared to perform. Decisions of when to close roads are based on years of experience. Agencies do not know the criteria for permitting or forbidding automation in adverse weather. Systems with differing functionality from various manufacturers have diverse abilities to handle adverse weather, and the abilities are certain to change over the years. Any new responsibilities for operating agencies will have to come with training and funding.

4.2 Weather-Related Limits of Automated Vehicles Are Unknown

Manufacturers are testing their vehicles in a variety of weather conditions. Limited experiments conducted for this project exemplified limits of the adverse weather conditions in which the driving automation features can perform. Owner's manuals acknowledge limits in weather but rarely give guidance on determining whether any particular conditions are in those limits. Consumers, and even engineers, have no way of knowing boundaries of operation for various functions.

A system operating at level 2 leaves responsibility with the human to determine whether conditions are safe. A system at level 3 or higher must confirm itself that it can handle the weather. It needs to know the conditions and to compare them with its operational design domain. Its own sensors will suffice in some situations (for example, when its vision system sees only rain). However, a vehicle performing at level 4 or 5 cannot assume that a human will be available to intervene. If it encounters adverse weather beyond its domain and no driver is available, it must enter a minimal risk condition, such as slowing or stopping. In heavy traffic, or especially if the passenger has special needs, this is not a satisfactory outcome. Therefore, prior information that a planned trip may pass through adverse weather and an assessment of whether that weather is within the operational design domain are essential.

Two inches of snow has a vastly different effect in Georgia than it does in Minnesota. A human driver may know that intuitively, but the actual surface conditions must be conveyed to an automated driving system.

5 Conclusion: Need for Adverse Weather Standards

The most significant gap is that there are no good ways of deciding whether a trip under automation should begin or continue. State and local operating agencies are ill equipped to give advice on automation use, and manufacturers are not advertising the limitations of their products. Currently, consumers are left to make the decisions. As automation features become more sophisticated, there needs to be an objective way to describe the current and forecast conditions so that they can be compared with the operational design domain of an automation system.

An approach is to develop a standard set of metrics. A manufacturer would define its operational design domain in terms of these metrics. The set might include visibility (for the driver's own vision and for all of the sensors on the vehicle), surface friction, local conditions (e.g. whether water is standing on the road or a tree has fallen across the road or a rolling work zone is repairing damage), and wind. If a manufacturer chooses to publish its capability, a human driver or dispatcher can assess the situation. If the exact limits are not published, the vehicle must be able to receive the forecast electronically, and then accept the trip, recommend against it, or outright refuse it. Manufacturers are an essential stakeholder in this process. Either individually or collectively (e.g., through an SAE committee), they must specify the nature of information needed to determine whether current and projected conditions are within the operational design domains of their systems.

FHWA, along with its State and local partner agencies, must collaborate with the National Weather Service, private sector weather providers, and the vehicle manufacturers at a system level to identify the tasks to be performed, determine which organization is best suited to perform those tasks, and carefully delineate interfaces between them. Existing standards and programs can be adapted, with gaps filled in to complete the system. Data would include current and forecast conditions for atmospheric weather and road weather, including spot warnings.

The weather and highway communities have a role in delivering information. Objective reports of conditions can come from conventional sources—the National Weather Service and road weather information systems. Road weather information currently coming from agency vehicles can be supplemented by information from civilian vehicles through already envisioned V2I communication.

The pieces of information will need to be checked for quality and assimilated into an overall picture, with the temporal and spatial specificity needed to support automated vehicle operations. These details need to be packaged in a form useful to the vehicles and their operators.

Vehicles that are making more than a short local trip need external sources of information on conditions over their planned route. They need to "see" conditions farther ahead, beyond the range of their own sensors. An open question is what organization or organizations will be responsible to collect, quality check, assimilate, and disseminate the information.

Acknowledgements. This work was funded by the Federal Highway Administration, Road Weather Management Program, under contract DTFH6116D000049 Order 0004.

References

Federal Highway Administration (FHWA): How do weather events impact roads?, February 1, 2017. https://ops.fhwa.dot.gov/weather/q1_roadimpact.htm. Accessed 18 Jan 2019

Hergeth, S., Lorenz, L., Krems, J.: Prior familiarization with takeover requests affects drivers' takeover performance and automation trust. Hum. Factors **59**, 457–470 (2017)

Koglbauer, I., Holzinger, J., Eichberger, A., Lex, C.: Drivers' interaction with adaptive cruise control on dry and snowy roads with various tire-road grip potentials. J. Adv. Transp. **2017**, 1–10 (2017)

Louw, T., Kountouriotis, G., Carsten, O., Merat, N.: Driver inattention during vehicle automation: how does driver engagement affect resumption of control. In: 4th International Conference on Driver Distraction and Inattention (2015)

Schwarz, C., Brown, T., Gaspar, J., Keum, C.: Transfer from highly automated to manual control: performance and trust. In: Proceedings of the 25th International Technical Conference on the Enhanced Safety of Vehicles (ESV) (2016)

SAE International: Work in progress. Requirements for V2I weather applications. J2945/3, January 17, 2012

SAE International: Surface vehicle recommended practice. Taxonomy and definitions for terms relating to driving automation systems for on-road motor vehicles. J3016, June 2018
Sundararajan, S., Zohdy, I.: Vehicle automation and weather: challenges and opportunities (No. FHWA-JPO-17-494), December 2016
United States Department of Transportation (USDOT): Prototype road weather performance management tool. FHWA-JPO-17-486, November 2016

Part IV Transportation Infrastructure and Planning

Designing Streets for Autonomous Vehicles

William Riggs[1(✉)], Melissa Ruhl[2], Caroline Rodier[3], and Will Baumgardner[2]

[1] University of San Francisco, 2130 Fulton Street, San Francisco, CA 94117, USA
wriggs@usfca.edu
[2] Arup, 560 Mission Street, Suite 700, San Francisco, CA 94105, USA
{Melissa.Ruhl,william.baumgardner}@arup.com
[3] UC Davis, 1605 Tilia Street, Davis, CA 95616, USA
cjrodier@ucdavis.edu

Abstract. How can automated vehicles be deployed on city streets to enhance urban and regional livability? This chapter outlines a visioning process where automakers, engineers, planning and policy professionals shared perspectives on how autonomous vehicles can be integrated onto city streets. It provides an engagement process as well as policy and design outcomes to help achieve aspirational streets of the future that promote equality of modes and environmental sustainability.

Keywords: Autonomous vehicles · Streets · Built environment · Cities · Urban planning

1 Introduction

Autonomous vehicles (AVs) stand to disrupt cities as much as they are likely to disrupt transportation systems [1]. Experts predict that vehicles accessible to the public in major cities could be fully automated within the next ten to forty [2]. The public sector is just beginning to understand automated vehicle technology and how it may interact with our land use and transportation systems. Automated vehicle technology could significantly reduce the time and cost of travel by car relative to other modes and, as a result, could increase congestion, vehicle miles traveled (VMT), greenhouse gas emissions (GHGs), and land consumption [3]. However, policies and plans at the street level offer significant opportunities to mitigate this advantage and establish rules that level the playing field for all modes, including shared vehicles, non-motorized, and transit.

The fastest growing and most invested-in transportation industries in recent years are networked and data driven personal travel—from companies offering variations of on demand services for travel via individual vehicles all from a networked, mobile platform. These companies connect riders with drivers through mobile smartphone apps, allowing point-to-point, ride-sourced travel [4]. These services include micromobility (e.g., e-scooter-sharing), microtransit (e.g., demand-responsive shuttles), ridesourcing (i.e., transportation network companies or TNCs), and a wide range of services yet to be popularized or invented. Because these services allow for convenient

G. Meyer and S. Beiker (Eds.): AUVSI 2019, LNMOB, pp. 111–122, 2019.
https://doi.org/10.1007/978-3-030-22933-7_12

mobility, it is possible that they reduce automobile ownership in urban areas with the potential to facilitate first and last mile connections [4, 5] and complement transit [6]. Yet most emergent data that arises from business as usual street policy and planning is indicating new, networked transportation, particularly vehicle-based services such as ridesourcing and microtransit, reduce public transit ridership and active travel [7], may increase VMT [8] and complicate curbside drop off [9].

This new mobility framework, as represented by TNCs and ridesourcing companies is a harbinger of autonomy. Many have explained the potential benefits of these vehicles. As visualized in Fig. 1, AVs also present new opportunities to connect individuals to jobs and change the way cities organize space and optimize trips. Particularly with AVs as shared or subscription-based services, the opportunities to reduce collisions, improve access to healthcare, and optimize emergency response [10–12] are significant. However, subscription or shared vehicle services could also increase emissions and congestions, depending on relative time and cost.

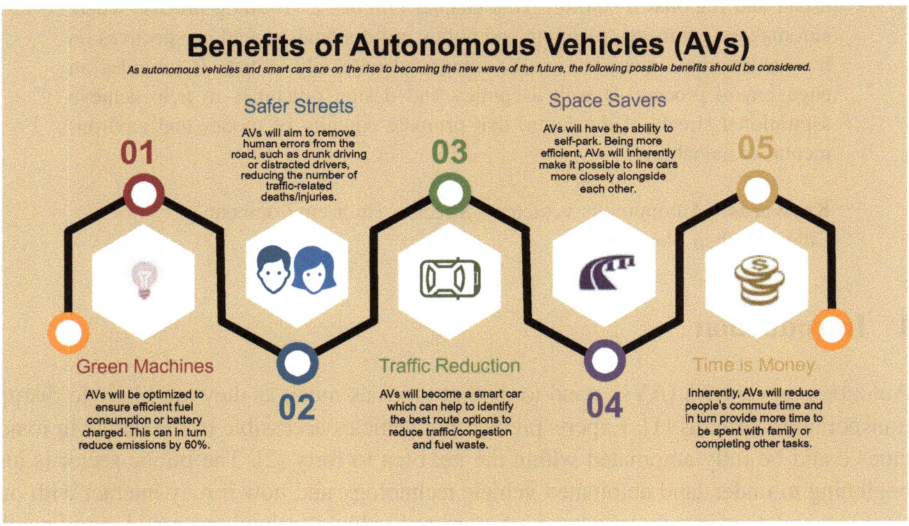

Fig. 1. The range of potential benefits of AVs discussed.

While some of these benefits are predicated on the extent to which car manufactures are able to transition, or perhaps wean, customers from the idea of owning a car, many car companies are working on a new kind of mobility that is more shared and multi-modal, and they are beginning to accept that the centrality of the personal automobile ownership in our public realm is outmoded. Mobility companies are welcoming the need for appropriate street design to reduce the risk of accidents due to conflicts between motorized and non-motorized travel and to increase the efficiency of curb-space use.

The importance of street system design on multi-modal accessibility that tempers growth in vehicle travel cannot be underscored enough. A large body of work discusses the importance street design has on sustainable travel behavior. This includes the ideas of connected networks, density thresholds and diversity of land uses [13, 14] as well as those of street configuration [15–17] and impacts of urban design features including sidewalks, trees, benches, and windows [18, 19–21].

Past work has dealt with these design implications and the secondary effects of automated vehicles on the built environment at wide scale [22]. Yet, there are also challenges and little understanding of how travel behavior and streets will change in this brave new world. Some work has used chauffer-based experiments to try to test behavior [23]—providing the real-world test of this technology thus far—and many other high-level policy documents have made suggestion and prediction that range in type and scale. Yet technological change is outpacing urban planning and policy.

Some of the most specific work combining data science takes incremental steps to envision future streets. These kind of tools which facilitate the participatory process have become an important part of the practice of planning and governance in recent years [27]. Tools that provide *interactive* and *transactive* platforms can help enrich community participation and democratize the policy-making process [28, 29]. In this chapter, we document the use of these tools to engage industry, planners, and policy makers in an AVS workshop entitled, "Integrating Automated Vehicles on City Streets." The goal was for participates to consider how they could transform the streetscapes to optimize inclusion of autonomous vehicles and create more livable communities.

2 Methodology

From a methodological perspective, the workshop was structured to move efficiently from dialogue to action. We asked participant to answer the question of how streets could be designed to facilitate autonomous travel and promote urban livability. The focus on the street enabled concrete design and policy exercises at different points of the development trajectory of fully automated vehicles. The framework for this process is illustrated in Fig. 2.

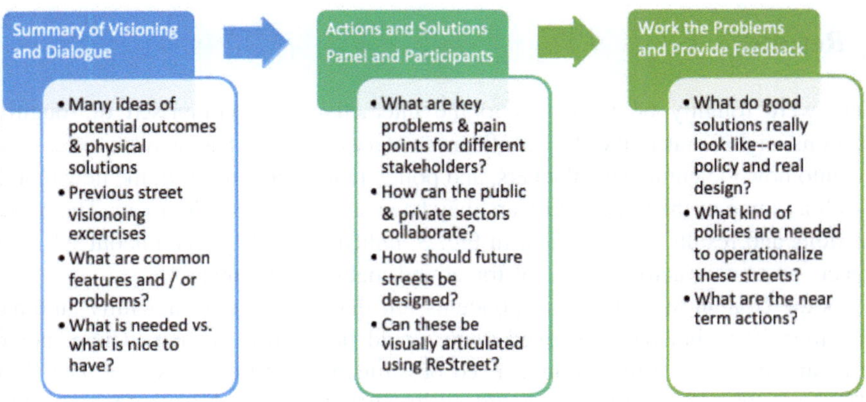

Fig. 2. Methodological framework.

The workshop began with an interactive poll and an introductory framing of travel behavior and street transformation. The goal was to provide a background on automated vehicles and the opportunities and challenges, that is travel behavior, design, economics, congestion and equity, of incorporating them into our current transportation system. Following this, an expert panel discussed potential pain points between cities and industry as they develop autonomous driving technology, exploring their priorities, goals and needs. The discussion was then used for the participants to act and engage in designing their own streets. This participatory design exercise was framed in 2 parts.

- Participants were challenged to develop consensus and write some of their own assumptions using the following prompt:

  Think about your assumptions for the future of mobility. Given what we've discussed in this session and the hopes /fears articulated, what do you think mobility will look like in 10-15 years for each of the 3 kinds of streets? What will the mix of vehicles be between automotive and other modes? Between AVs and non-AVs?

- They were then challenged to sketched out potential concepts for three 'subject' streets using the ReStreet planning tool [24, 25, 30]. ReStreet is a digital participatory planning application (app.restreet.com). It allows for users to change the default street width, the building height, the number of lanes and type of lanes on a given street. They can then submit this future vision of the street online where the data can be synthesize for policy and decision-making. The prompt provided read:

  *Using the restreet app (*http://app.restreet.com*) develop conceptual street designs for each of the assumed scenarios, or use the paper in front of you to sketch the new street. Limit yourself to no more than 2 streets per table (we want you to make tradeoffs!) and make sure you submit your street using the dropdown on the right! (If you have to make policy /program trade-offs document them!)*

After this process, the new street visions were presented, and then critiqued by the expert panel, who offered concluding thoughts as well as policy and design takeaways.

3 Results

There were roughly 66 responses to the interactive poll, comprised of roughly 6 questions, significant at the 90th confidence interval $\pm 10\%$. The answers offer a window into how the engineers, planners, and policy makers in a room frame new mobility as well as some of the hopes and fears they have about their development. The primary questions and results are presented in Fig. 3, indicating a skepticism about AVs, their impacts on transit, active modes of travel and metropolitan areas.

As the data shows 86% of respondents felt that fleets would not fully automated until after 2040, but most believed there would be interim impacts. These included optimism about active modes and mixed pessimism about transit. As the data shows, 56% of respondents said transit would decline while 40% said it would grow. About 4% said there would be no change in transit use at all.

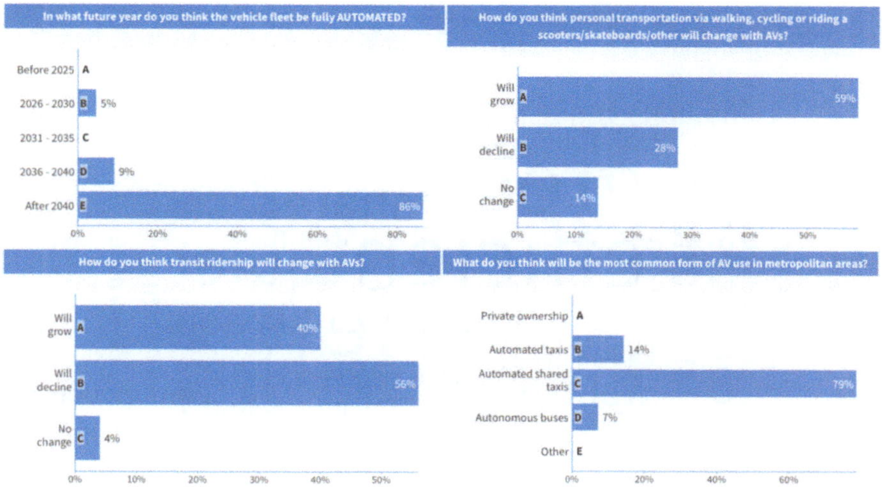

Fig. 3. Interactive survey questions about AV development.

Likewise, the vast majority (79%) thought automated vehicles would be shared, despite prior presentations on the challenges of enabling such a path. Despite this, when asked what their hopes for automated vehicles, many people used words to indicate efficiency, safety and environmentally conscious travel. The opposite was true when asked their fears. As shown in Fig. 4, the top keyword for what people fear about future streets with AVs is "congestion."

Given this baseline, participants embarked on a modeling exercise to design streets for an automated future. They first wrote out assumptions. They then designed streets using the ReStreet tool. Two of the redesigned streets that were submitted are illustrated in Fig. 5.

Key themes of these redesigned streets included that they were people-centric and legible. Consistent with the work of Schlossberg and others [26] most individuals used principles of parking removal, roadway thinning and repurposing of street right of way to support pedestrian and non-automotive mobility infrastructure. Some noted that the street should be able to support other uses, including housing and that the tool did not allow for that.

Finally, building on these street visions workshop participants then outlined the policies needed to shape these kinds of streets. These policies are summarized by theme in Table 1 for commercial corridors. As the table indicates, a constant policy theme is zoning of the street and allocation of space. This relates to speeds and modal priorities but also to parking and curb use. Also, there are clear and specific design suggestions for policy including multiple suggestions for wider sidewalks, eliminating dangerous left turns and integrating freight /logistics/deliveries.

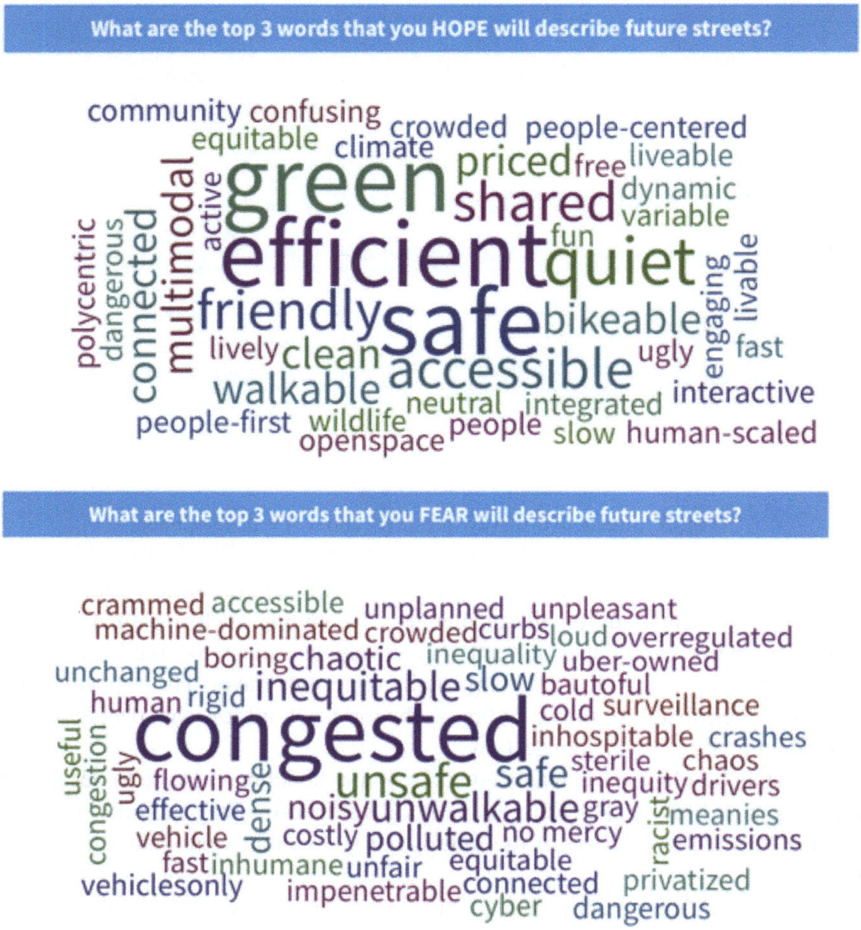

Fig. 4. Interactive survey questions about AV development.

Some of the same policy objectives hold for other street typologies at the neighborhood and arterial level, with a heightened importance on reducing or better controlling speed. As illustrated in Table 2, there was a strong emphasis on speeding and throttle control, with discussion of the concept from the NACTO Blueprint for Autonomous Urbanism that “twenty is plenty” (or a 20-mph speed limit on city streets). There were also interesting ideas about delivery that warrant further study, including pricing delivery (of parcels or people) based on proximity to the destination. Delivery to the neighborhood or the block should be cheaper than delivery to the door and policy might be developed to support these kind of pricing/shared mobility frameworks so they are consistent with livability and placemaking goals.

Fig. 5. Visions of future streets using ReStreet.

Table 1. Commercial Corridor Street Assumptions, Attributes & Policies

Land use	Mode/types of mobility	Traffic volume	Percent of AV/fleet mix	Pick up/drop off experience	Policies	Other policy/design notes
Urban	Transit; shared	High	Shared AV	Cut outs at every corner	*Limited access: shared transit use (bus/shuttle/shared ride) *Electric/low speed *Curb control (access) *Speed table	*Increase sidewalk width; 10-foot-wide sidewalks +4 ft for tree/greenspace; separated lanes by speed; 1 lane in each direction
High density urban	15% bike; 35% pedestrian; 35% transit; 5% scooter; 10% other	High	75% legacy; 25% AVs (bus, private, car)	Off-loading for passengers	*Low to no emissions with phase out (incentivizing) - no vehicle traffic except for loading *Reservation system for freight *Only drop off and pickup for loading/unloading *Shared standards *Remove parking in favor of loading	*10-foot sidewalks both sides; new loading zone *10-foot bike lane

(*continued*)

Table 1. (*continued*)

Land use	Mode/types of mobility	Traffic volume	Percent of AV/fleet mix	Pick up/drop off experience	Policies	Other policy/design notes
Urban	25% bike; 35% ped; 35% transit	High	50% AV	People and goods delivery only	*Delivery policy *Infrastructure standards *Eliminate left turn lanes	*Bus boarding middle of street *Flex or business zone *Staged delivery
High density urban	Heavy pedestrian traffic in the evening (after 6 pm)	High	Flexible dedicated/closed space	Still need freight/goods delivery	*Zones: low speed zones, EV-only zones; no personal vehicles (Denver model) *Curb management and use fees, parking, pickup/drop off. Charging *Flexible use *Curb bulb outs to helps AVs see pedestrians crossing *Delineation by mode (scooter wars) *Delineation by speed (bikes in center with through vehicles) *Enforcement	*Widen walkable space to more than 10 ft *Elevated crosswalks *Replace parking lane with pickup/drop-off area *Delineate parking vs passenger zones *Center transit lane for AV so it can be narrower, similar to BRT *Bikes high speed throughout

Table 2. Neighborhood and arterial corridor street assumptions, attributes & policies

Land use	Mode/types of mobility	Traffic volume	Percent of AV/fleet Mix	Pick Up/drop Off experience	Policies	Other Policy/design notes
Suburban	Bikes; personal vehicles	Low	More AV delivery	Delivery	Eliminate left turn lanes because future AV algorithms can avoid left turns	
Suburban	60% vehicles; 20% PT; 20% NMT	Low	50% legacy; 50% SAV	Priced by location - highest at door or community choice	EU directive: 20 kph residential streets: twenty is plenty	One loading/parking lanes on each side of street; one driving lane needed for low speed low ADT
Suburban	60% SOVs; 20% private cars; 20%					

bike/ped/transit/scooter/othersLowPeople and goods delivery onlyBreak up parking/loading with vegetation to limit through traffic and high speeds

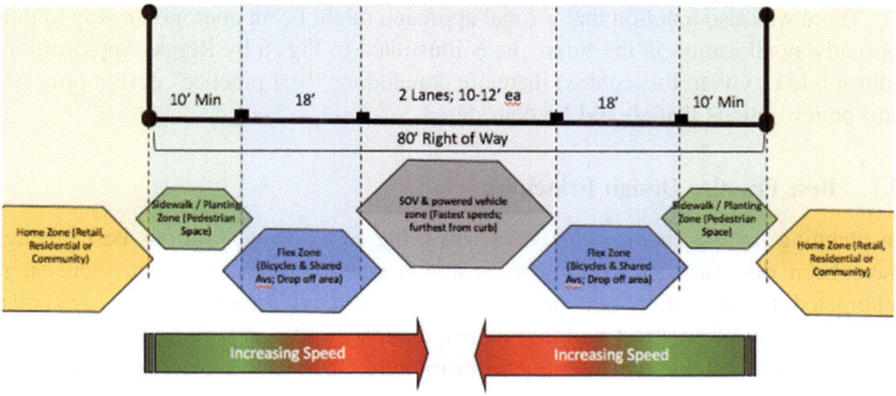

Fig. 6. A zonal approach to street space allocation in an autonomous future.

4 Conclusions

Autonomous vehicles are continually discussed and touted as the next big disruptor to society as we know it. They are expected to change the way we travel, reduce greenhouse gas emissions, eliminate congestion, and make mobility more accessible to a greater number of people. Detractors of this technology, however, argue that AVs will likely have the opposite effects – that AVs will increase congestion, increase GHG emissions, increase congestion, and cause mobility to be less equitable. While it is difficult to determine the exact impacts, consequences, and timing of this technology, it is clear that it will result in changes. One such area where these changes may first be seen in on the infrastructure serving this technology – our roadways.

What should streets look like in an autonomous future? Key findings and lessons learned from this process posed the idea that cities need to evolve from auto-centric to people-centric, and that streets should reflect this reality. There was consensus, even from autonomous vehicle manufacturers that, that particularly on city streets AVs need enable safe travel for people traveling on all modes. Potential new visions for both suburban and urban streets involved streets that were legible for vehicles but also that reframed a large component of street estate for transit users, cyclists, and pedestrians.

Additional policies that arose included the following key observations, including those on energy, curb management and design for vulnerable populations as bulleted in more depth below.

- Transportation and the electric grid will be intertwined. We need to plan for renewable energy systems for mobility.
- Today, the curb is critical; tomorrow, the curb may be irrelevant with new flexible designs that are not yet possible today.
- Accessible, universal design is a challenge for both vehicles and streets. This challenge needs to be tackled and resolved.

There was also a notion that a zonal approach might be an appropriate way to think of roadway allocation in the future, as is illustrated in Fig. 6 by Riggs, Appleyard and Johnson [31]. Given this context there are concluding "best practice" design principles and policy actions that should be considered.

4.1 Best Practice Design Principles

In planning for the changes that AVs will bring to roadways, various best practices have been discussed in working to ensure a smooth transition to an increasingly automated transportation network. These are clustered around 4 basic principles: reduce amount of on-street parking; reduce roadway width; repurpose the streets with pedestrian and alternative mobility infrastructure; look at roadway technology primarily for electrification.

First, as AVs become more widely used and available, private vehicle ownership is expected to decrease in favor of transportation as a service (TaaS)—meaning that instead of private ownership, individuals will pay for rides as they are needed. With TaaS, vehicles can be in operation 24/7 so instead of sitting parked 95% of the time, as most vehicles are, they would be more efficient and continually serving consumers. As a result, the need for on-street parking would decrease.

Second, in addition to the reduction in on-street parking, AVs will also remove the human error in driving and reduce the number of collisions and accidents. With reduced human error, vehicle travel lanes can be reduced. and thereby ensure that lane widths can be decreased. Reduced lane widths would provide additional space that could be reused or repurposed for various functions including those described below.

Third, as this chapter shows repurposing streets with bike lanes, sidewalks, and other features can support pedestrian-friendly environments and alternative support other methods of mobility with the additional space from the reduced right-of-way. With the additional space, new infrastructure supporting pedestrian and alternative mobility options could be developed.

Finally, the transition from existing roadway infrastructure to infrastructure that accommodates fully-automated vehicles will not be immediate and is likely to take significant time, investment, and coordination to achieve. In other words, developing infrastructure that accommodates AVs will occur incrementally and there is a need for minor changes over time. This may include smart technology that is installed within roadways to capture real-time data, but more importantly it needs to support trends toward electrification.

4.2 Policy Actions

With regard to policy one of the clear findings from this work, and outcomes from this chapter, is the needs to have a more agile and open policy dialogue. There is a dramatic need for public and private professionals to interact and explore more collaboration and partnerships—some of which can be facilitated with technology. The example of the tool used, ReStreet, provides a good illustration of this. Built as a partnership with Code for America, it is a technological tool that has been used to assist with the planning process. By allowing for private citizens and business owners to deliberate online and in-person, it facilitated citizen compromise and tradeoffs—important and oft missing components of citizen engagement.

Talking and partnering on street design is just the start. It may likely lead to partnerships in mapping, digital infrastructure and pricing—all of which form key opportunities and challenges for the road ahead.

References

1. Riggs, W.: Disruptive Transport: Driverless Cars, Transport Innovation and the Sustainable City of Tomorrow. Routledge, London (2019)
2. Underwood, S.E.: Disruptive innovation on the path to sustainable mobility: creating a roadmap for road transportation in the United States. In Road Vehicle Automation, pp. 157–168. Springer, Cham (2014)
3. Wadud, Z., MacKenzie, D., Leiby, P.: Help or hindrance? The travel, energy and carbon impacts of highly automated vehicles. Transp. Res. Part A: Policy Pract. **86**, 1–18 (2016)
4. Rayle, L., Dai, D., Chan, N., Cervero, R., Shaheen, S.: Just a better taxi? A survey-based comparison of taxis, transit, and ridesourcing services in San Francisco. Transp. Policy **45**, 168–178 (2016)
5. Shaheen, S., Chan, N.: Mobility and the sharing economy: potential to facilitate the first-and last-mile public transit connections. Built Environ. **42**(4), 573–588 (2016)
6. Rassman, C.L.: Regulating rideshare without stifling innovation: examining the drivers, the insurance gap, and why Pennsylvania should get on board. Pitt. J. Tech. L. Pol'y **15**, 81 (2014)
7. Gehrke, S.R., Felix, A., Reardon, T.G.: Substitution of ride-hailing services for more sustainable travel options in the greater Boston region. Transp. Res. Rec., January 2019. https://doi.org/10.1177/0361198118821903
8. Clewlow, R.R., Mishra, G.S.: Disruptive transportation: the adoption, utilization, and impacts of ride-hailing in the United States. Research Report UCD-ITS-RR-17-07. University of California, Institute of Transportation Studies, Davis, CA (2017)
9. Riggs, W.W., Boswell, M.R., Zoepf, S.: A new policy agenda for autonomous vehicles: it's time to lead innovation. Planetizen (2017)
10. Fagnant, D.J., Kockelman, K.M.: The travel and environmental implications of shared autonomous vehicles, using agent-based model scenarios. Transp. Res. Part C: Emerg. Technol. **40**, 1–13 (2014)
11. Guerra, E.: Planning for cars that drive themselves metropolitan planning organizations, regional transportation plans, and autonomous vehicles. J. Plann. Educ. Res., November 2015. https://doi.org/10.1177/0739456X15613591
12. Boswell, M., Riggs, W.: Could autonomous vehicles save lives in disasters? - Features | Planetizen. Planetizen (2017). https://www.planetizen.com/features/95813-could-autonomous-vehicles-save-lives-disasters. Accessed 11 Feb 2019
13. Ewing, R., Cervero, R.: Travel and the built environment – a meta-analysis. J. Am. Plann. Assoc. **76**, 265–294 (2010)
14. Cervero, R., Kockelman, K.: Travel demand and the 3Ds: density, diversity, and design. Transpo. Res. Part D: Transp. Environ. **2**(3), 199–219 (1997)
15. Gayah, V., Daganzo, C.: Analytical capacity comparison of one-way and two-way signalized street networks. Transpo. Res. Rec.: J. Transpo. Res. Board **2301**, 76–85 (2012)
16. Riggs, W., Gilderbloom, J.: Two-way street conversion: evidence of increased livability in Louisville. J. Plan. Educ. Res. **36**(1), 105–118 (2016)

17. Riggs, W., "Hans" Gilderbloom, J.I.: How multi-lane, one-way street design shapes neighbourhood life: collisions, crime and community. Local Environ. **22**(1), 1–17 (2017)
18. Riggs, W., Gilderbloom, J.: The connection between neighborhood walkability and life longevity in a midsized city. Focus **13**(1), 11 (2016)
19. Hajrasouliha, A., Yin, L.: The impact of street network connectivity on pedestrian volume. Urban Stud., August 2014. https://doi.org/10.1177/0042098014544763
20. Southworth, M.: Designing the walkable city. J. Urban Plan. Dev. **131**, 246 (2005)
21. Ewing, R., Hajrasouliha, A., Neckerman, K.M., Purciel-Hill, M., Greene, W.: Streetscape features related to pedestrian activity. J. Plann. Educ. Res., July 2015. https://doi.org/10.1177/0739456X15591585
22. Riggs, W., Larco, N., Tierney, G., Ruhl, M., Karlin-Resnick, J., Rodier, C.: Autonomous vehicles and the built environment: exploring the impacts on different urban contexts. In: Meyer, G., Beiker, S. (eds.) Road Vehicle Automation 5, pp. 221–232. Springer, Cham (2019)
23. Harb, M., Xiao, Y., Circella, G., Mokhtarian, P., Walker, J.: Projecting travelers into a world of self-driving cars: naturalistic experiment for travel behavior implications. In: Proceedings of the 97th Transportation Research Board. Washington D.C. (2018)
24. Boswell, M.R., Riggs, W., Ross, R.: Streetplan: hacking streetmix for community-based outreach on the future of streets. Focus: J. Plann. Pract. Educ. | J. City Reg. Plann. Depart. **13**, 59 (2016)
25. Riggs, W.W.: Modeling Future Street Options in an AV Future Using Restreet (2017)
26. Schlossberg, M., Riggs, W.W., Millard-Ball, A., Shay, E.: Rethinking the street in an era of driverless cars. Urbanism Next (2018)
27. Evans-Cowley, J., Griffin, G.: Microparticipation with social media for community engagement in transportation planning. Transp. Res. Rec.: J. Transp. Res. Board **2307**, 90–98 (2012)
28. Evans-Cowley, J., Kubinski, B.: A Brave New World: How Apps Are Changing Planning. Planetizen: The Urban Planning, Design, and Development Network, September 2012. http://www.planetizen.com/node/58314. Accessed 16 Dec 2014
29. Riggs, W., Gordon, K.: How is mobile technology changing city planning? Developing a taxonomy for the future. Environ. Plann. B Plann. Des., October 2015. https://doi.org/10.1177/0265813515610337
30. Riggs, W.: Technology, civic engagement and street science: hacking the future of participatory street design in the era of self-driving cars. In: Proceedings of the 19th Annual International Conference on Digital Government Research: Governance in the Data Age, pp. 4:1–4:6. Delft, The Netherlands (2018)
31. Riggs, W., Appleyard, B., Johnson, M.: A design framework for livable streets in the era of autonomous vehicles. In: Proceedings of the 98th Annual Meeting Transportation Research Board. Washington, D.C. (2019)

Building Automation into Urban and Metropolitan Mobility Planning

Bart van Arem[1(✉)], Aernout Aki Ackerman[2], Tilly Chang[3], William Riggs[4], Augustin Wegscheider[5], Scott Smith[6], and Siegfried Rupprecht[7]

[1] Faculty of Civil Engineering and Geosciences, Delft University of Technology, Postbus 5048, 2600 GA Delft, The Netherlands
b.vanarem@tudelft.nl

[2] MABS Consultancy, Damplein 3, 2265AC Leidschendam, The Netherlands
aki@mabsconsultancy.nl

[3] San Francisco County Transportation Authority, 1455 Market Street, 22nd Floor, San Francisco, CA 94103, USA
tilly.chang@sfcta.org

[4] University of San Francisco, 2130 Fulton Street, San Francisco, CA 94117, USA
wriggs@usfca.edu

[5] Boston Consulting Group, 300 N LaSalle St, 46th Floor, Chicago, IL 60654, USA
wegscheider.augustin@bcg.com

[6] U.S. Department of Transportation Volpe National Transportation Systems Center, 55 Broadway, Cambridge, MA 02142, USA
Scott.smith@dot.gov

[7] Rupprecht Consult – Forschung & Beratung GmbH, Clever Str. 13-15, 50668 Köln (Cologne), Germany
s.rupprecht@rupprecht-consult.eu

Abstract. Transport authorities and mobility planning stakeholders have started discussing approaches to planning for road automation in cities and metropolitan areas. We present guiding principles for developing AV ready mobility plans, and show how scenario development, travel demand and transport modelling and participatory street redesign can be part of a planning approach at different temporal and spatial scales. We show best practices from the Rotterdam-The Hague, Boston and San Francisco Bay areas. We recommend a multi-stakeholder approach, integrating AV based travel with other options such as transit, TNC, walking and cycling and embedding transport with other fields such as energy, land use and equity.

Keywords: Urban and Metropolitan Mobility Planning · Scenario development · Transport modelling · Automation readiness

G. Meyer and S. Beiker (Eds.): AUVSI 2019, LNMOB, pp. 123–136, 2019.
https://doi.org/10.1007/978-3-030-22933-7_13

1 Introduction: Planning for Automation

Transport authorities and mobility planning stakeholders have started discussing approaches to planning for road automation in cities and metropolitan areas. Although various uncertainties will continue to exist for many years, planners in many countries aim to facilitate the deployment of automation concepts that are contributing to achieving mobility policy goals; in addition, there is a growing awareness that infrastructure built today should be as compatible as possible with future needs of automated mobility. Different approaches are being pursued on the international scale. In 2018, US DOT convened the FHWA National Dialogue [1], with several workshops including one on planning and policy. Issues considered include automation impacts on travel demand, land use, infrastructure investment, right of way use, policy barriers. FHWA is leaning towards scenario techniques, with an interest in a national concept of operations.

The EU's approach to 'Sustainable Urban Mobility Planning' (SUMP) [2] that is based on an agreed policy concept [3] is currently being updated to include also new policy areas such as automation. The following SUMP mobility planning principles are particularly relevant for automation and will be important criteria for creating 'automation-ready' cities and metropolitan regions [4]:

(1) Aim for sustainable (automated) mobility! Any policy or (technology deployment) that is included in a SUMP should contribute to a more sustainable mobility system. Therefore, it is essential to establish concrete concepts of how automation will contribute to solving major urban challenges such as improved economic, environmental, social development, better management of urban space, improved safety (also for vulnerable road users), socially acceptable levels of cybersecurity that meet citizens' privacy concerns.

(2) Cover the entire 'functional city'! One of the strongest impacts of automation will be on land use patterns in the wider (metropolitan) region. Therefore, it will be particularly important to plan across municipal boundaries and to develop strategies for balancing impacts throughout the entire 'functional city'.

(3) Develop a long-term vision of automation! Scenario techniques are recommended to better manage uncertainty on the 'big disruption challenges', e.g. acceptance of business models, changing legal frameworks, dynamics of vehicle deployment. Stakeholders participating in mobility planning should then try to understand the most likely interactions and dynamics for their community to decide what actions might be taken immediately, in five or in ten years [5].

(4) Assess the expected performance of an automated transport system! 'What are key challenges that automation needs to solve in the functional city?' As tools (e.g., for modelling) and data are becoming available, planners will be increasingly able to establish more specific automation impacts. Other, more generic questions about the wider 'transport system' are: Which skills are planning organizations missing? Which institutional structures need updating? Which revenue streams are we losing (e.g. reduced parking fees), what do we need to compensate for losses? Which policies/regulations make automation less desirable ("policy audit").

(5) Cooperate widely: across institutional boundaries, involve citizens and stakeholders! Automation is part of a fundamental shift in the organization of urban transport, requiring a wide range of stakeholders to become part of the required change process. Planners need to consider: Who are relevant old and new actors, what is the new playbook (e.g. energy suppliers, OEMs, IT startups, real estate)? Who benefits from what? How can we contribute to a realistic perception and image of automated vehicles among the general public and among decision makers? What do we need from state and federal/national government to regulate?
(6) Arrange for monitoring and evaluation! What would be automation specific 'SMART targets'.[1] How can they be measured? How could a dynamic system of monitoring and decision-making increase policy? Who should collect, own such performance data, who should be able to create revenues from it?
(7) Develop an action plan for all (automated) transport modes! Automation provides the possibility of a new collective mobility paradigm, consisting of shared vehicle use, on demand (automated) last/first mile services, integrated mainstream transit. However, it is unclear how current mainstream public transit operators will be able to adapt to new business models and how cooperation between traditional and new market actors can work in practice. Planners should identify opportunities and facilitate the exploitation of synergies to help create a more effective transport system.

Overall, transport planning authorities should aim to become 'automation ready' in the sense that they should develop a (flexible) planning approach, based on SUMP principles and focused on delivering public policy goals [6]. An example of such an 'automation readiness' plan might include the following elements (Fig. 1):

2 Methodologies and Tools

2.1 Scenario Development and Analysis

The advent of Automated Vehicles challenges regional and urban authorities on how to take AVs into account in their long-term decision making. In this process several uncertainties need to be dealt with: at which rate and with which functionality will AVs be deployed. Will shared AVs reduce the need for parking space? Will AVs complement or compete with public transport? Will AVs reduce traffic congestion? Will it be safer? Will infrastructure have to be adapted?

In a comprehensive review Milakis et al. [7] classify the implications of AVs into 3 ripples: (1) traffic, travel cost and travel choices, (2) vehicle, infrastructure and location choice and land use and (3) energy, air quality, safety, social equity, public health. The majority of a growing body of literature is still focusing on the implications in the first – inner - ripple, while research into the wider implications is in its infancy. In order to prepare for AVs as well as electric and shared means of mobility, Babes [8] developed

[1] 'SMART targets' are Specific, Measurable, Assignable, Realistic, Time-related.

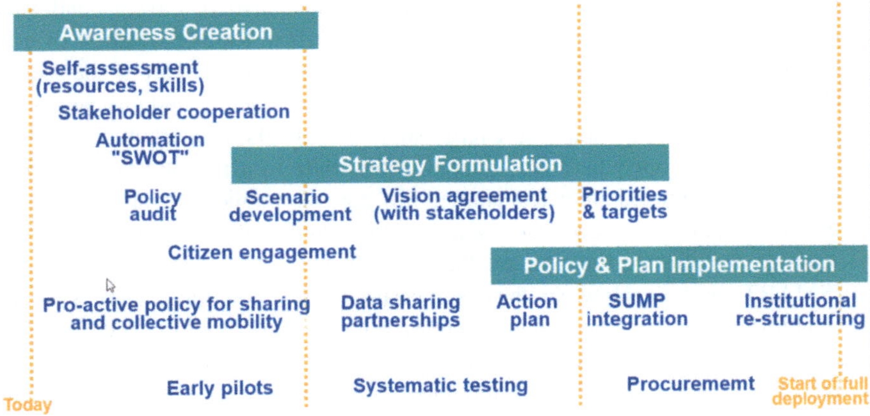

Fig. 1. Example of an 'automation readiness' plan

alternative urban and regional designs using separation of flows and shared space as design principles (Fig. 2).

As a first step to assess the feasibility and implications of a spatial and mobility systems incorporating AVs, Milakis et al. [9] used a Delphi approach to develop scenarios how AVs could be part of mobility in 2030 and 2050, based on alternative assumptions for technology maturity as well as government support. Nieuwenhuijsen et al. [10] use a system dynamic model to explore the diffusion of AVs in the 'AV in Bloom' scenario [9] assuming high technology maturity and strong governmental support. In this scenario, increasing levels of automation replace the lower levels as they are becoming available, with L1, L2 and L3 reaching fleet penetration levels around 10%, 25% and 50% between 2025–2030.

As a second step, the mobility impacts are assessed as function of AV penetration rate. Puylaert [11] used continued on the use of system dynamics in [B4]. Mobility impacts where modeled in 6 months increments through a classic static equilibrium multimodal transport model on a simplified functional road network of the Netherlands. AVs were assumed to make car driving more attractive by reducing the Value of Travel Time (VOTT) by 80–90% [12] and make traffic more efficient by assigning passenger car unit (PCU) values of 95% and 85% to autonomous and cooperative AVs, respectively [13]. The results in [11] suggest a wide range in possible impacts, including the generation of car demand in and between large cities and potentially leading to massive 'AV congestion'. Madadi et al. [14] looked at the impacts on mobility of only equipping a subnetwork for L4 automation assuming specific infrastructural provision such as markings, connectivity and separation from other traffic at higher speeds. Assuming a static equilibrium transport model with a lower VOTT and lower PCU values, impacts on mobility show AV drivers changing routes towards AV equipped roads, leading to lower total cost at an increase in total travel time.

Finally, the impacts of AVs were modeled as a sensitivity analysis using the Dutch National Transport Modelling system (LMS) [15]. The LMS is used in a 4 year

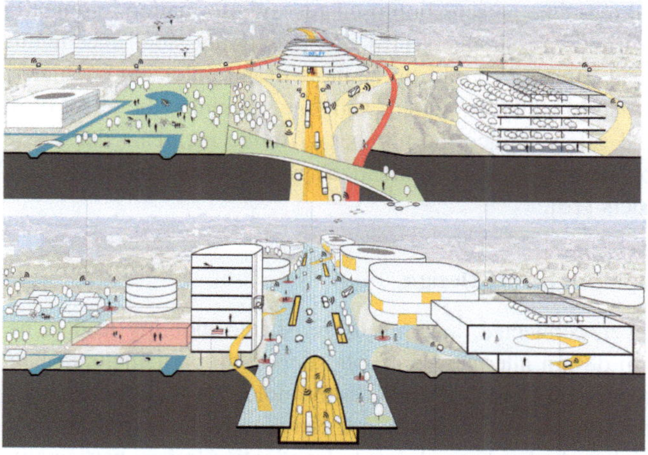

Fig. 2. Redesign A13 Rotterdam The Hague Motorway using AVs [8].

planning cycle to support long term transport infrastructure investment strategies in the Netherlands. Scenarios were defined based on variations of VOTT and PCU values for truck platooning, autonomous and cooperative passenger cars. In particular for the scenarios using cooperative AVs, congestion is reduced compared with a reference scenario without any AV. Decreasing the VOTT leads to an increase of travel distance and travel time at the cost of reintroducing congestion.

2.2 Travel Demand Modeling

The program of the 2018 TRB Innovations in Travel Modeling (ITM) Conference [16] noted that "With the dawn of a new era of transformative disruptions in transportation – characterized by automation and connectivity, autonomy and sharing, and on-demand mobility and delivery – the need for innovations in travel modeling has never been greater. Transport modelers are being challenged every day to explore new frontiers in big data analytics and harness the power of machine and deep learning algorithms to simulate and forecast mobility demand and travel patterns under a wide range of future scenarios of automation and connectivity." While automation was only a small part of the 2016 conference, it became much more significant in 2018. One speaker noted that it is an unsettling time to be a modeler, because fundamental assumptions (e.g., regarding travel mode and value-of-time) are being challenged by AVs. The social and policy context, including land use, urban form and traveler attitudes, needs to be considered.

There is increased interest in exploratory modeling [17]. On the demand side, uncertainties around AVs include automobile ownership (privately owned AVs vs. shared services), the level of car and ride sharing, empty vehicle trips, value of time, induced trips and changes in parking location and behavior. Components of empty vehicle trips include fleet repositioning, intra-family repositioning (e.g., the family car returning home to serve a second family member), and repositioning to avoid expensive parking. One presentation used survey information to quantify the disutility from sharing trips with strangers [18]. On the supply side, uncertainties include the changes in road capacity caused by connected and automated vehicles (C/AV) [19], C/AV infrastructure including signals and lane markings, the effect of traffic incidents (will improved safety reduce the number of traffic incidents that cause delay?), and shared fleet operations. Table 1 summarizes presentations that included AVs in existing planning models [16].

2.3 Participatory Street Design

At a street level, connected, automated and shared mobility lead to new options for street and curb design alternatives. Tools supporting participatory planning processes have arisen as a result of mobile and GIS technology and are becoming more available to citizens for participation planning and community development [20]. For example, tools like social media has been shown to be effective 'micro-participation' in planning processes and provide an important part of the practice of planning and governance [21]. Tools that provide interactive and transactive platforms can help enrich community participation and democratize the policy-making process [22]. Transactive tools involve an exchange between agencies and citizens, conducting business that had previously required an in-person presence an opportunity to be conducted on-line. Applications that have allowed for crowdsourcing and user input from a larger community contributing to a larger body of information include public input tools like Mindmixer. Economic development platforms like SeeClickFix or GoRequest, allow citizens to directly report issues in their neighborhood to their local governments so they can be addressed by a local agency.

The Restreet.com (app.restreet.com) tool can help facilitate participatory street design for new mobility and automated /autonomous vehicles [Fig. 3, 23–25]. The tool was designed to democratize street planning and synthesize that data for policy and decision-making. The tool was used to show right-of-way needs eroding due to the prevalence of autonomous vehicles creating efficiency and to dialogue the policy decisions needed to do so in advance of their widespread adoption.

In the pilots the team set up a bank of computers with wifi-enabled hotspots for the public to congregate and submit their street plans. The process of having citizens work through design options alongside others is very useful. People engage in important discussions on trade-offs (e.g. if you have more space dedicated to bicycles then that leaves less space not only for cars but for pedestrians), and make decisions based on these discussions; grappling with these issues in parallel with others. First experiences of using the tool in a participatory environment, let to a consensus that in a shared AV environment, it might be possible to remove on-street parking, thin lanes, and think are as a future vision for streets [26–28].

Table 1. Planning model applications using AVs; hyperlinks in [16]

Planning modeling AV application	Location
Incorporating Connected and Autonomous Vehicles and Ride-Hailing Services in the Traditional Four Step Model	Dallas/Ft.Worth, Texas
Activity-Based Model with Dynamic Traffic Simulation to Explore Scenarios for Private and Shared Autonomous Vehicle Use	Jacksonville, Florida
Modeling Connected and Automated Vehicles in a Four-Step Travel Demand Model	Chittenden County, Vermont
A Framework for Modeling Connected and Autonomous Vehicles in the New Statewide Model	Michigan
Development and Application of a Model to Estimate Driverless Autonomous Vehicle Trips	Minneapolis/St. Paul, Minnesota
New Mobility: Integrating Autonomous Vehicles, the Sharing Economy and Impacts of E-Commerce into a Model Framework	Vancouver, British Columbia
Shared Autonomous Vehicles as a Replacement for Buses: A Simulation Study	Berlin
Estimating Potential Impacts of AVs and TNCs using an Activity-Based Travel Demand Model in MTP/SCS Scenario Development	Sacramento, California
Impact Assessment of Autonomous DRT Systems	Berlin
Autonomous Vehicle (AV) Scenarios using a VisionEval model	Oregon

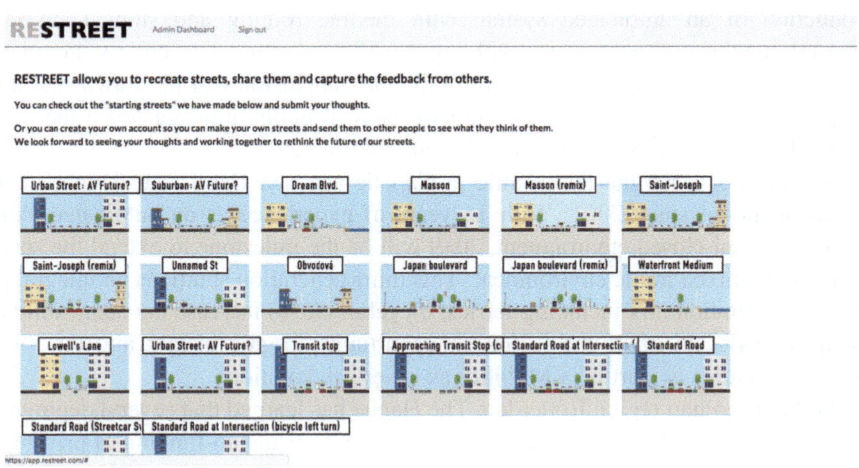

Fig. 3. ReStreet interface

3 Lessons Learnt in Cities

3.1 Automated Transport Metropolitan Region of Rotterdam the Hague

The Innovation Network AVLM (Automated Transport on the Last Mile) in the Metropolitan region of Rotterdam and The Hague is built on the premise that automated transport allows the region to develop better public transport services against lesser cost with an increase in travelers using the public transport systems. Furthermore, the implementation of automated transport systems enhances the economic potential of the region. The goal is to develop multiple AVLM projects to gain insights in the economic potential of automated transport and forge consortia between public offices, solution developers, transport service providers, and knowledge institutions to find ways to exploit the opportunities. Currently, ten locations are under various stages of development.

The Research Lab Automated Driving Delft has been established in 2017 to allow for first stage, closed track, testing. Semi closed testing has been allowed on the TU Delft Campus. The other nine locations shall be used for open road testing in real-life environments. With a fleet of automated vehicles available to private and public partnerships, the focus is on gain knowledge and experience in terms of human vehicle interaction, acceptance, and user experience.

The location Leidschendam-Voorburg, as cooperation between the municipality and a private real-estate developer, is a good example of the alignment of private and public benefits sought from automated transport. Where the focus of the real-estate developer is on securing the accessibility for visitors of the Mall of the Netherlands, the municipality aims to scale to a system that allows for an ageing population to have access to suitable mobility solutions. Noteworthy is that the transport analysis showed that the introduction of automated vehicles on a fixed track with regular stops, increased the number of passengers using public transport by 14%, Whereas the introduction of an automated system with flexible routing and stops showed an increased number of passengers with up to 50%. Among the elderly population increases up to 67% were shown. From this, the conclusion can be drawn that automated transport system enhance the attractiveness of public transport once the transition can be made to flexible, demand driven systems [29].

Last, but not least, the Parkshuttle in Capelle aan den IJssel near Rotterdam has been in operation since 2001. With 15 years of experience as an automated shuttle system in a semi-closed environment, 2019 will be the milestone to extend the service into an open, mixed traffic environment. This third-generation shuttle is an integral part of the transportation mix including ferries, bicycles, etc. to increase the accessibility of a business park and offers the region an opportunity to gain insight into the role of automated shuttles as part of a Mobility as a Service solution.

The Metropolitan region Rotterdam The Hague has learned that several deployment locations are needed for learning, along with cooperation and co-funding of private and public stakeholders. This accelerates learning as knowledge is quickly and easily transferred to, and implemented at, other locations. Multiple locations are also needed to explore the different aspects of the business case behind automated transport so that it true economic value can be established and exploited to create sustainable, cost efficient, accessible public transport systems.

3.2 City of Boston

In July 2016, the World Economic Forum (WEF) and Boston Consulting Group (BCG) partnered with the City of Boston to assess the impact of AVs in the city, to catalyze testing of AVs there and to strategize how the city could foster this technology to achieve its mobility goals. In 2017, a large-scale conjoint analysis was conducted to forecast the penetration of several types of AVs in Boston's future modal mix. Research participants were presented with variables, such as the length of the trip and the time of day, and were required to make discrete choices about what mode of transport they would use. This approach generated a realistic and granular view of how mobility will evolve in Boston [30].

The analysis predicts a clear shift to mobility-on-demand (for both autonomous and traditional vehicles), which will account for nearly 30% of all trips in the Greater Boston area and 40% of trips within city limits in the future. Driving this shift are the cost-competitive nature of robo-taxis and robo-shuttles – especially on shorter trips – and the added convenience and comfort compared with mass transit. In suburban and other areas outside the city, the analysis showed that mobility-on-demand will mainly replace personal-car usage. In urban areas, it will replace the use of both personal cars and mass transit, to equal degrees, with the shift creating a risk of increased congestion.

To understand the effects of AVs in Boston, a traffic simulation model was built that showed the contrasts between current traffic patterns and future scenarios. The simulation used the results of the conjoint analysis – including the projected modal mix of personal vehicles, taxis, private AVs and shared AVs. Three important findings emerged: (1) Shared AVs will reduce the number of vehicles on the streets and reduce overall travel times across the city. Findings showed that the number of vehicles on the road will decrease by 15% while the total number of miles travelled will increase by 16%. However, travel time will improve by just 4% on average – not as dramatic as other studies have forecast, but still an improvement. (2) Introducing shared AVs will worsen congestion in the downtown area, mostly because these vehicles will serve be chosen as substitutes for short public transportation trips. Travel time will increase by 5.5% in downtown Boston. In Allston, a neighborhood outside the city's core, mobility-on-demand will mainly replace the use of personal cars rather than mass transit, and travel time will decrease by 12.1%. (3) With the new modal mix, Boston will require roughly half as many parking spots, including those on streets and in parking structures. AVs present an opportunity to rethink the overall design of the city's streets.

Local governments hold the key to influencing these results because they have the power to implement the right policies and incentives. The greatest effects are likely to come from occupancy-based pricing schemes, in which financial incentives discourage single-occupancy rides. This measure could improve citywide travel time by 15%. When Boston formally announced its collaboration with the Forum in September 2016 no companies were testing AVs in the city. After just four months and executive orders at both city and state level, AV operator nuTonomy completed the first autonomous mile in Boston in January of 2017, and by November AVs from the partnership of nuTonomy and Lyft, which provided the booking platform, began travelling with passengers on board. By the end of the year, Optimus Ride and Aptiv were also participating in the pilot, and more than 1,500 autonomous miles had been completed in an expanded testing area. Further expansion of testing continues in 2018, with Optimus Ride's approval for passenger trials counting as one recent achievement.

The key to success in Boston was collaborative leadership. The mayor, along with his transportation team and their counterparts in the Massachusetts state government, committed to embracing a transformational technology, taking advantage of Boston's highly innovative workforce in software and robotics. State and local government worked hand in hand with the participating AV operators to test options, learn, iterate and scale. Other cities can follow this model to introduce AVs onto their roads and successfully collaborate with the private sector.

3.3 San Francisco Bay Area

In the San Francisco Bay area regional and local governments are beginning context-sensitive planning for Autonomous Vehicles. Led by Arup and sponsored by the Association of Bay Area Governments (ABAG) and Metropolitan Transportation Commission (MTC), the Horizon initiative explores AV strategies for the region. This is important because these new technologies have not been addressed by current long-range planning processes. The planning developed numerous perspectives and a "toolbox" strategies for consideration in future planning documents—in this case called Plan Bay Area 2050 [31, p. 201]. Unique to this project the goal for the Bay Area was that the goal for AVs should be consistent with regional goals and guiding principles:

- Affordable (and tied to affordable housing)
- Connected (with intercity and local transportation options that help create a cohesive region)
- Diverse (supporting people of all backgrounds, abilities and ages)
- Healthy (support public health and environmental goals, where open space, clean water and clean air are preserved)
- Vibrant (be synergistic with regional innovation and economic growth efforts.

The study examined potential impacts of automation with respect to these guiding principles, and developed several strategies including repurposing of parking, control of greenfield development, continued support of high-capacity transit corridors, suburban automated demand-responsive micro-transit, dynamic road pricing, data sharing, as well as attention to equity, safety and workforce concerns. It puts policy objectives forward including dynamically priced and demand responsive transit, along with incentives to increased shared use of vehicles.

Specifically in San Francisco, agencies (SFCTA, SFMTA) are working on several areas to respond to a myriad new mobility transportation services operating in San Francisco. These areas include identifying goals, policy frameworks [31], permit programs and legislation (a per-trip ride-hail fee will go before voters in 2019), Transit First and Vision Zero initiatives to ensure equitable access, and coordinating advocacy on state and Federal rulemakings. Data collection and research are important in this early period of experimentation, to measure progress toward goals such as the efficient movement of people and goods versus vehicles and to inform and shape new policies such as curb management and congestion pricing. This work requires a fresh commitment of public resources to developing professional capacity and analysis tools.

One example is San Francisco County Transportation Authority's recent TNC's Today research series, which seeks to quantify and characterize ride-hail activity in San

Francisco. The SFCTA is particularly interested in this subject because TNCs (Transportation Network Companies) might be viewed as an AV 1.0, with attendant positive and negative impacts, requiring policies to support efficient and equitable outcomes (reliable transit and HOVs) as well as public infrastructure provision, management and regulation. Recent studies estimate that TNCs account for up to 1 in 4 vehicles in the downtown core and about 50% of the rise in congestion between 2010–2016, with the balance caused by population and employment growth [32]. The SFCTA is also a sponsor of new mobility services, designing a Federally funded automated shuttle service to support a new development on Treasure Island, to be deployed as a first/last mile connection to the Island's transit system. A related project adds an integrated trip planning and payment system to make it easy for passengers to book shuttle transfers to San Francisco and East Bay ferries and buses.

4 Discussion

Worldwide, metropolitan and city mobility planning authorities have responded to the challenge of integrating new forms of mobility emerging based on technological innovations regarding connectivity and automation into mobility planning. At the US federal level as well as at EU level, programs and guidelines have been developed, not only building on best practices of front runner (member) states, counties and cities, but also empowering other metropolitan and city authorities to prepare and make the best use of connected and automated mobility.

Recent research has resulted in a collection of design and modelling tools to explore AV introduction pathways as well as first order impacts on mobility. Initial ranges for changes in roadway capacity and Value of Travel Time are available. At a local level, initial research methods and results are becoming available on alternative roadway and street design, based on participatory planning and making use of new mobility options in order to improve local livability. Real-world experiments with AVs in urban and regional setting will provide a more solid basis to reduce current uncertainties in current model applications Further attention is needed is to understand the wider societal implications, in particular in combination with urban and regional spatial planning.

First experiences of metropolitan and city authorities with integrating connected and automated mobility into their planning processes, show that planning is now being challenged by differing time scales. Technology is evolving quickly, while long-range plans are on a 20–30 year time scale, and infrastructure is on a ~100 year timescale. The "classic" modeling process is based on quantifiable elements (e.g., travel time), and leads to a point forecast. Given the uncertainties in the long-range impacts of automation, and the important factors such as quality of life, public health and equity that may be more difficult to quantify, it may be time to ask whether the "classic" process is still working. New approaches include adaptive policy making, scenario exploration and learning from pilot projects.

New challenges are clearly emerging in the field of street redesign in relation to parking and curb space allocation. C/AVs as well as TNC services may change the demand for parking, create a need for pick-up/drop-off points and also open up possibility to improve local livability. Guiding principles are being developed for pilots, localization, safety, data (define metrics) and the ability to scale and maintain. Immediate actions include defining the problem, engagement with stakeholders, curbside pilots with TNCs (perhaps at airports), and acting early to ensure that future options, such as infrastructure changes, are kept open.

Many uncertainties exist around the requirements road operators need to take into account in roadway operation and maintenance. It is expected that roads would will need to be prepared for a mix of CAVs and traditional manned vehicles during an interim period. Performance should be measured in people rather than vehicles per hour. There may be a need for electric charging infrastructure, as well as dedicated AV infrastructure (HOV lane conversion, platooning on long trips).

C/AV mobility is also part of future services by new mobility providers such as Uber, Waymo and Lyft. Metropolitan and city transport planners clearly see a role for TNCs, although there is a need to set guiding principles. Adaptability is key, and such principles will have to ensure multiple options (including shared, automated electric vehicles), prioritize transit, ensure equity, and consider pricing levels to address congestion, the environment, and urban sprawl.

5 Conclusions

Modelers, planners and policy makers are actively pursuing automation as a mobility option. There is a growing consensus about the direction of opportunities and challenges. Opportunities include the potential of improved mobility for non-motorists while a challenge is increased travel, resulting in increased congestion. Shared mobility and last mile transport are priority areas. Common questions relate to the new mix of shared/on-demand, mainstream collective and personal ownership-based transport (and cycling and walking). Uncertainties about deployment, user response, and impacts are major barriers to concrete planning. First best practices of tools and methods are available, but are lacking data from real-world applications. Pilots provide a basis to verify assumptions and collect data.

Based on the research, analysis and stakeholder discussion reported in this Chapter, we recommend that planning for automation should be data-based and policy-guided. Scenario planning and other exploratory analysis can help to handle uncertainty, but more work on methodologies and use of new data sources is needed to really understand the mutual impacts of AVs on travel behavior, mobility, transport networks and regions. Wide collaboration with stakeholders and new actors is a key requirement for planning, considering also links to energy, land use, environment, and social inclusion. In the present era of dawn of the deployment of AVs, we strongly recommend continuation of an international dialogue on ‘automation policy principles’.

References

1. FHWA: Factsheet National Dialogue on Highway Automation (2018). https://ops.fhwa.dot.gov/automationdialogue
2. Rupprecht Consult – Forschung und Beratung: Guidelines. Developing and Implementing a Sustainable Urban Mobility Plan (2014). http://www.eltis.org/sites/default/files/sump_guidelines_en.pdf
3. European Commission: COM (2013) 913 Final. Annex 1: A Concept for Sustainable Urban Mobility Plans to the Communication From The Commission 'Together towards competitive and resource-efficient urban mobility' (2013). https://ec.europa.eu/transport/sites/transport/files/themes/urban/doc/ump/com(2013)913-annex_en.pdf
4. Heinrichs, D., Rupprecht, S., Smith, S.: Making automation work for cities: impacts and policy responses. In: Meyer, G., Beiker, S. (eds.) Road Vehicle Automation 5. Cham (2018)
5. Gyergyay, B., Gomari, S., Friedrich, M., Olstam, J., Rupprecht, S.: Automation-ready framework for urban transport and infrastructure planning. 25th ITS World Congress, Copenhagen (2018)
6. Rupprecht, S., Backhaus, W., Gyergyay, B., Gomari, S.: CoEXist Deliverable D1.1 Automation Ready Framework (2018). https://www.h2020-coexist.eu/wp-content/uploads/2018/12/D1.1-Automation-Ready-Framework-Preliminary-version-1.pdf
7. Milakis, D., van Arem, B., van Wee, B.: Policy and society related implications of automated driving: A review of literature and directions for future research. J. Intell. Transp. Syst. Technol. Plann. Oper. **21**(4), 324–348 (2017). https://doi.org/10.1080/15472450.2017.1291351
8. Babes, V.: A Method to Assess the Impact of Automated Vehicles on Urban Liveability in the Rotterdam The Hague Metropolitan Region. European Post-Master in Urbanism (EMU), P5 report (2017). repository.tudelft.nl
9. Milakis, D., Snelder, M., Van Arem, B., Van Wee, B., De Almeida Correia, G.H.: Development and transport implications of automated vehicles in the Netherlands: Scenarios for 2030 and 2050. Eur. J. Transp. Infrastruct. Res. **17**(1), 63–85 (2017)
10. Nieuwenhuijsen, J., Correia, G.H.D.A., Milakis, D., van Arem, B., van Daalen, E.: Towards a quantitative method to analyze the long-term innovation diffusion of automated vehicles technology using system dynamics. Transp. Res. Part C Emerg. Technol. **86**, 300–327 (2018). https://doi.org/10.1016/j.trc.2017.11.016
11. Puylaert, S., Snelder, M., van Nes, R., van Arem, B.: Mobility impacts of early forms of automated driving – a system dynamic approach. Transp. Policy **72**, 171–179 (2018). https://doi.org/10.1016/j.tranpol.2018.02.013
12. Correia, G.H.D.A., Looff, E., van Cranenburgh, S., Snelder, M., van Arem, B.: On the impact of vehicle automation on the value of travel time while performing work and leisure activities in a car: Theoretical insights and results from a stated preference survey. Transp. Res. Part A Policy Pract. **119**, 359–382 (2019). https://doi.org/10.1016/j.tra.2018.11.016
13. Hoogendoorn, R., Van Arem, B., Hoogendoorn, S.: Automated driving, traffic flow efficiency, and human factors. Transp. Res. Rec. **2422**, 113–120 (2014). https://doi.org/10.3141/2422-13
14. Madadi, B., van Nes, R., Snelder, M., van Arem, B.: Assessing the travel impacts of subnetworks for automated driving: An exploratory study. Case Stud. Transp. Policy **7**(1), 48–56 (2018). https://doi.org/10.1016/j.cstp.2018.11.006

15. Smit, R., van Mourik, H., Verroen, E., Pieters, M., Bakker, D., Snelder, M.: Will self-driving cars impact the long term investment strategy for the Dutch national trunk road system? In: Proceedings European Transport Conference (2017). https://aetransport.org/en-gb/past-etc-papers/conference-papers-2017
16. Transportation Research Board: 7th Innovations in Travel Modeling Conference, Atlanta, Georgia, USA (2018). http://onlinepubs.trb.org/onlinepubs/conferences/2018/ITM/Program.pdf
17. Sun, S., Stabler, B., Milkovits, M.: Risk analysis workshop. Presented at the 7th Innovations in Travel Modeling Conference, Atlanta (2018). http://onlinepubs.trb.org/onlinepubs/Conferences/2018/ITM/workshop2a.pdf
18. Lavieri, P.S., Bhat, C.R.: Modeling individuals' willingness-to-share trips with strangers in an autonomous vehicle future. Presented at the 7th Innovations in Travel Modeling Conference, Atlanta (2018). http://onlinepubs.trb.org/onlinepubs/Conferences/2018/ITM/PLavieri.pdf
19. Li, X.: CAV Trajectory optimization & capacity analysis - modeling methods and field experiments. Presented at the 7th Innovations in travel Modeling Conference, Atlanta (2018). http://onlinepubs.trb.org/onlinepubs/Conferences/2018/ITM/XLi.pdf
20. Slotterback, C.S.: Planners' perspectives on using technology in participatory processes. Environ. Plan. **38**(3), 468–485 (2011)
21. Evans-Cowley, J., Griffin, G.: Microparticipation with social media for community engagement in transportation planning. Transp. Res. Rec. **2307**, 90–98 (2012)
22. Riggs, W., Gordon, K.: How is mobile technology changing city planning? Developing a taxonomy for the future. Environ. Plann. B **44**(1), 100–119 (2017). https://doi.org/10.1177/0265813515610337
23. Riggs, W.: Modeling Future Street Options in an AV Future Using Restreet (2017)
24. Riggs, W.: Technology, civic engagement and street science: hacking the future of participatory street design in the era of self-driving cars. In: Proceedings 19th Annual International Conference on Digital Government Research: Governance in the Data Age, pp. 4:1–4:6. Delft, Netherlands (2018)
25. Riggs, W., Boswell, M.R., Ross, R.: Streetplan: hacking Streetmix for community-based outreach on the future of streets. Focus J. Plann. Pract. Educ. J. City Reg. **13**(1), 14 (2016)
26. Schlossberg, M., Riggs, W.W., Millard-Ball, A., Shay, E.: Rethinking the street in an era of driverless cars. UrbanismNext (2018)
27. Appleyard, B., Riggs, W.: 'Doing the right things' before "doing things right": a conceptual transportation/land use framework for livability, sustainability, and equity in the era of autonomous vehicles. Presented at the Transportation Research Board 97th Annual Meeting, Washington, DC (2018)
28. Riggs, W., Appleyard, B., Johnson, M.: A design framework for livable streets in the era of autonomous vehicles. In: Proceedings of the 98th Annual Meeting Transportation Research Board. Washington, DC (2019)
29. Ackerman, A.A.: Haalbaarheidsstudie AVLM Gemeente Leidschendam-Voorburg (in Dutch, Feasibility Study ATLM City Leidschendam-Voorburg). MABS Consultancy (2018)
30. World Economic Forum: Reshaping Urban Mobility with Autonomous Vehicles - Lessons from the City of Boston. REF 140518 (2018). http://www3.weforum.org/docs/WEF_Reshaping_Urban_Mobility_with_Autonomous_Vehicles_2018.pdf. Accessed Jan 2019
31. Metropolitan Transportation Commission: Autonomous vehicles perspective paper, June 2018. https://mtc.ca.gov/sites/default/files/2018-06-25_Autonomous_Vehicles_Perspective_Paper.pdf. Accessed 29 Jan 2019
32. San Francisco Transit County Authority: Emerging Mobility Factsheet, November 2017. https://www.sfcta.org/emerging-mobility. Accessed 29 Jan 2019

Digital Infrastructure for National AV-Readiness

Jennifer Carter[1] and Valerie Shuman[2(✉)]

[1] Intelligent Transportation, HERE Technologies, 425 W Randolph Street, Chicago, IL 60606, USA
jennifer.carter@here.com

[2] SCG, LLC, 5141 Crain Street, Suite 300, Skokie, IL 60077, USA
vs@shumangroupllc.com

Abstract. We know that we need to build a public/private digital infrastructure to support a coast-to-coast, AV-ready transportation environment – but what does that term actually mean? And what key definitions and models do we need to provide a common vocabulary and roadmap so that agencies can each build their piece of this future system, which must transcend jurisdictional borders?

This session was an opportunity for global stakeholders to hear from the experts on current progress and collaborate to: develop a common definition of "AV-ready digital infrastructure", generate a short-list of required definitions and models; and review existing frameworks, including a proposed Data Maturity Model.

Keywords: Digital infrastructure · AV-ready · Cooperative transportation · Cooperative automated vehicles · CAV

1 Introduction

The 2018 Automated Vehicle Symposium, organized by the Association for Unmanned Vehicle Systems International and the Transportation Research Board, included a Breakout Session on *What's a Digital Infrastructure, Anyway? Building A Shared Vision for National AV-Readiness*. This session included two Rocket Roundtables, in which expert panelists shared their perspectives about the following questions:

- What is a digital infrastructure for transportation?
- What are the key components /capability areas of a digital infrastructure?
- How will our digital infrastructure make us "AV-Ready" (what should it look like to enable cooperative, automated transportation)?
- What are the top three national definitions/models that must be in place to enable development of the digital infrastructure?
- What might an actual model look like, e.g., data maturity?

After each Roundtable, over forty participants from around the world collaborated in Jam Sessions to produce their own answers. This chapter provides a summary of these discussions. It is intended to serve as a basis for further work in defining,

G. Meyer and S. Beiker (Eds.): AUVSI 2019, LNMOB, pp. 137–142, 2019.
https://doi.org/10.1007/978-3-030-22933-7_14

designing and implementing the national digital infrastructure which is needed to enable next-generation transportation solutions.

2 Session Topic Detail

We know that we need to build a public/private digital infrastructure to support a coast-to-coast, AV-ready transportation environment – but what does that term actually mean? And what key definitions and models are needed to provide a common vocabulary and roadmap so that agencies can each build their piece of this future system, while ensuring interoperability across jurisdictional borders? Breakout Session #19 organized this discussion into two key topics:

2.1 Topic #1: Definitions

This segment was an opportunity to collaborate on a common definition for the term "digital infrastructure". John Corbin, Connected Automated Vehicle Program Manager, FHWA Office of Operations; Tracy Larkin-Thomason, Deputy Director of Nevada DOT Southern Nevada; and Serge Van Dam, Principal Advisor Traffic Management, Rijkswaterstaat Ministry of Infrastructure and Water Management shared their ideas in a discussion moderated by Jennifer Carter, Sr. Manager, Government Solutions, HERE Technologies. Highlights included:

- The challenge is talking broadly enough. We need a compelling national vision.
- A digital infrastructure is a set of foundational services to support transfer of data, which in turn supports efficiency
- We need to:
  - address new approaches to physical infrastructure management, planning, education, and funding
  - look at coordinating local, state and national standards as well as national regulatory frameworks so that our overall system works across jurisdictional boundaries
  - protect resources such as spectrum which are required to support critical activities like emergency response
  - address the practical aspects of sharing and data exchange; existing processes systems are not always well suited for this
  - remember the diversity of time frames when planning – some transportation system decisions will continue to last for decades
  - be realistic about what's possible given diverging stakeholder perspectives; e.g., are the OEMs really ready to start listening to their environment? Can they really depend on external data and services as part of their customer offering?
  - figure out appropriate roles and responsibilities in this new environment
- It is time to revisit our priorities. What is the set of transportation data and services that are currently relevant?

2.2 Topic #2: Key Frameworks

This segment focused on generating a list of key definitions and models needed to enable continent-level implementation, and collect specific feedback on emerging work in this area. Sandra Larson, Transportation Innovation Strategies Leader, Stanley Consultants, moderated a panel which included Scott Marler, Director, Operations Bureau, Iowa DOT; Dr. Jon Neff, Aerospace Corporation/Civil Systems Group, Artificial Intelligence Team, Center for Space Policy on Autonomous Systems; Nikola Ivanov, PMP, Deputy Director, University of Maryland Center for Advanced Transportation Technology Laboratory; and Valerie Shuman, Principal, SCG, LLC. Highlights included:

- We should consider learning from approach taken by the Internet Engineering Task Force (IETF), which has cooperatively developed a set of key interface standards which have very effectively supported dramatic evolution over time. This process might be adjusted for the CAV environment to include some government-mandated standards.
- A digital infrastructure is an ecosystem that enables a seamless flow of information
- We need to learn from Amazon – their operations move a lot of cardboard boxes, but the real value is in their digital infrastructure.
- The digital infrastructure centers on data and we need to focus on: the operational environment, archiving and predictive capabilities, resilience and equity.
- We should think carefully about what will really motivate the development of our new digital infrastructure – what's driving this work?
- We need to develop national-level frameworks that will allow both humans and machines to effectively use data as they move throughout our transportation ecosystem; data maturity and quality are a key part of this.

3 Outcomes

At the end of each Jam Session, each table reported out to the entire group. After the event, a further review of the detailed notes from the speakers and each table's insights revealed the following Big Ideas:

3.1 Moonshot Vision

There was general consensus that we are at the beginning of a new chapter in transportation, and that we need to define our vision for what's next. We need to respect existing systems and processes (like planning, construction, operations and maintenance) while integrating a whole new suite of capabilities which will help optimize the capabilities of our overall system. In particular, we need to:

- *Extend our definition of "physical infrastructure"* to include the data lifecycle support systems which will be needed to handle the expected influx of massive amounts of transportation-related data. This includes computing, storage, communications, etc.

- *Add a "digital infrastructure"* which can enable a continuously evolving set of data-driven transportation services and capabilities to serve both public and private consumers. Both the data and the users will be highly diverse, and will require real-time solutions to everything from daily mode choice to emergency response.
- Recognize that *both physical and digital components will be highly integrated*, requiring the removal of many existing silos. Data will need to be able to move freely and be processed effectively throughout the system.

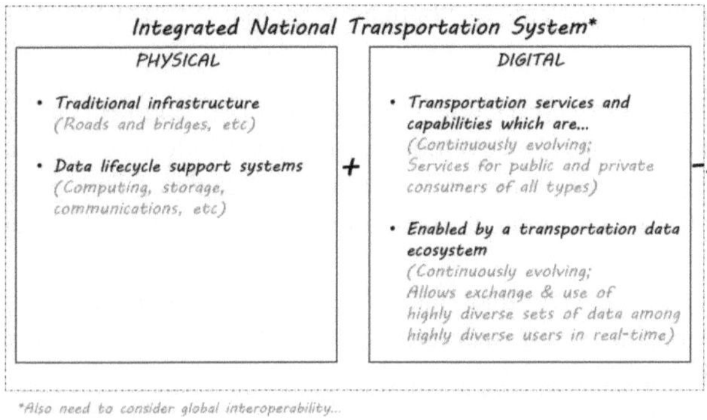

Fig. 1. Integrated national transportation system

Figure 1 shows what this new integrated National Transportation System vision might look like.

3.2 Digital Infrastructure Definition

Attendees put forward quite a number of different definitions for the term "digital infrastructure". Reviewed as a group, many of these definitions had concepts in common, which might be summarized as:

A digital infrastructure is a transportation data ecosystem
governed by a set of institutional policies and technical standards

It is particularly important to recognize that a digital infrastructure:

- *Combines data, interfaces, and physical infrastructure.* We need to make appropriate investments to support the entire stack. The term "IT modernization" was used repeatedly in this context.
- *Is not comprised of technology alone.* Development of key policies and standards is critical to successful development of the ecosystem. One group proposed a "data exchange model" which included common vocabulary, reference architecture, hierarchical models and digital rights models.

Figure 2 captures the details behind the definition provided by the session participants.

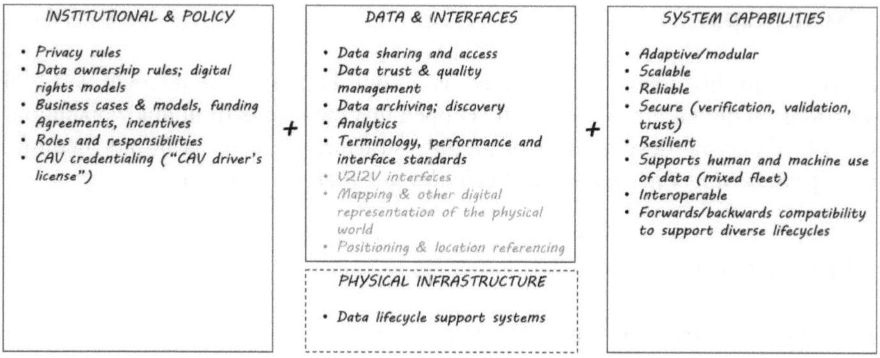

Fig. 2. Digital infrastructure definition

3.3 Practicalities

Many of the participants in the discussion are front-line practitioners with a great deal of direct experience in building and operating transportation solutions. As a result, the group was quite clear that we need a very practical approach to next steps. In particular:

- *We must assume continuous change.* This concept was repeated many times in the discussion. While many participants naturally saw this as a big challenge, a particularly telling quote was: "We will be in transition. Deal with it".
- *It is time to formalize some of the critical rules, standards, and interfaces; and to implement those changes.* There are many experiments and new programs underway which have already started to address pieces of the digital infrastructure puzzle, and we now need to put a larger framework in place.
- *We can't fail to work on the system-level items if we intend to make our moonshot.* However, it might make sense to tackle a prioritized set of capabilities as a way to maintain momentum, such as a limited set of national public sector data "products" or the top three priority data exchanges.
- *There are important opportunities to learn from others.* Possible sources of insight included the IETF approach to building and maintaining internet standards, existing transportation asset management processes; and the aviation, aerospace and electrical industries.

3.4 Frameworks and Models

The group also spent some time reviewing a strawman data quality model which was developed by the Iowa AV program, building on work conducted by the European ITS Platform (EIP+) Organization. Opinions on the model were quite diverse, reflecting the many different perspectives in the room. It was clear that a great deal of further discussion will be needed as we work to prioritize and define appropriate models of this type.

4 Next Steps

The content in this document is offered as a contribution to the national and global discussions on digital infrastructure. It is hoped that the results from this workshop will help to facilitate the development of the infrastructure we need in furtherance of our shared goal: ***getting fully automated vehicles out there saving lives.***

5 Further Resources

Participants noted the following additional resources:

- National Dialogue on Highway Automation discussion on Digital Infrastructure https://ops.fhwa.dot.gov/automationdialogue/
- U.S. DOT work Data for AV Integration: https://www.transportation.gov/av/data
- SAE ORAD committee is working on industry vocabulary in its J3131 effort
- The Connected Vehicle Pooled Fund Study has been gathering industry vocabulary in a Glossary: http://www.cts.virginia.edu/wp-content/uploads/2018/03/Glossary-of-CAV-Terms-Ver1.0-03052018-1.pdf.

Acknowledgements. The authors wish to thank the other members of organizing team for this breakout session: John Corbin, Transportation Specialist, FHWA; Sandra Larson, Transportation Innovation Strategies Leader, Stanley Consultants; and Dennis Paul, Senior Project Leader, The Aerospace Corporation. To establish an open forum for frank discussion, it was agreed that individual input to this workshop would be unattributed. However, the organizers do wish to thank the panelists and attendees who gave so generously of their time and ideas.

Author Index

G. Meyer and S. Beiker (Eds.): AUVSI 2019, LNMOB, pp. 143–144, 2019.
https://doi.org/10.1007/978-3-030-22933-7

MIX
Papier aus verantwortungsvollen Quellen
Paper from responsible sources
FSC® C105338
FSC
www.fsc.org

If you have any concerns about our products,
you can contact us on
ProductSafety@springernature.com

In case Publisher is established outside the EU,
the EU authorized representative is:
Springer Nature Customer Service Center GmbH
Europaplatz 3, 69115 Heidelberg, Germany

Printed by Libri Plureos GmbH
in Hamburg, Germany